C&EN $27.50

ADAPTIVE RADAR
IN REMOTE SENSING

ADAPTIVE RADAR IN REMOTE SENSING

BY
DAG T. GJESSING

ANN ARBOR SCIENCE
PUBLISHERS INC / THE BUTTERWORTH GROUP

Copyright © 1981 by Ann Arbor Science Publishers, Inc.
230 Collingwood, P. O. Box 1425, Ann Arbor, Michigan 48106

Library of Congress Card Catalog Number 81-68032
ISBN 0-250-40487-7

Manufactured in the United States of America
All Rights Reserved

Butterworths, Ltd., Borough Green, Sevenoaks, Kent TN15 8PH, England

PREFACE

We are indeed living in a decade of privileges: radio sciences have progressed to the stage where much of the initial hard work has successfully been completed, leaving a foundation from which new achievements can be made, permitting realization of complex solutions not hitherto conceivable. The radio science theoretician is furnished with a powerful set of tools in the form of flexible computers and a library of software for a wide variety of fundamental problems. We can draw on our predecessors, who developed much of the basic methodology and provided basic knowledge.

The experimentalist is privileged to have a set of components from which a diagnostic tool can be structured. An example is the tunable laser. During the last two decades, the spectral intensity of illumination sources (ultraviolet, visible or infrared light) has been improved by something like six orders of magnitude. In recent years, superheterodyne optical receivers that can be matched to the illuminator have been developed, providing a sensitivity improvement which is also something like six orders of magnitude. For certain experiments, we have hence improved the overall remote sensing capability by something like 12 orders of magnitude. Few generations of scientists have been exposed to such opportunities. Objects or phenomena with hitherto unidentifiable properties, even when confined to the laboratory, can now be studied from remote platforms.

Progress in the field of microwave techniques is equally breathtaking. The preceding generation of scientists had little more for microwave illumination than magnetrons, whose frequencies were largely fixed and determined by the mechanical geometry. Today we are privileged to have solid-state microwave sources which can be controlled in amplitude and phase (frequency) by microprocessors in a predescribed manner, thus providing us with illuminators which can be matched (adapted) to the object of interest.

High-speed digital processors enable us to implement sophisticated signal processing and pattern recognition algorithms in real time.

New technology emerging from solid-state physics based on the interaction between acoustical waves in piezoelectric materials and electromagnetic waves is about to open another set of avenues in regard to high-speed signal analysis.

In addition to being blessed with challenging advances in basic science and engineering, we are living in a decade where our achievements in science and engineering can be transformed into relevant and stimulating applications. In the history of science, never has the time span between basic research and mission-oriented applications been as short as today.

In addition to serving the more "classical" fields of applied research, such as radio communications, and military and space research, radio methods now have established their potential in environmental science. This is largely a result of an upsurge of interest in problems concerned with the earth's resources, pollution and conservation. This has led to a general need for a more detailed understanding of the governing principles in environmental systems and living organisms. The time has now come when the two fields are merging, resulting in proliferating progress in radio science applications.

We, as radio scientists, obviously face a very challenging future: that of developing remote sensing systems for the surveillance of Mother Nature, pollution on the negative side, resources on the positive. A very wide field of disciplines is involved, ranging from physics and chemistry through geology, geophysics and biology to ecology. Obviously, our ambition cannot be to measure all of the variables that play a role in this very complicated control system. Our endeavor must be through dialogs with the environmental scientist to select a set of observable properties that adequately describes the environmental system. It then remains to measure these state parameters with adequate resolution in space and time in accordance with the established rules of information theory (Figure 1).

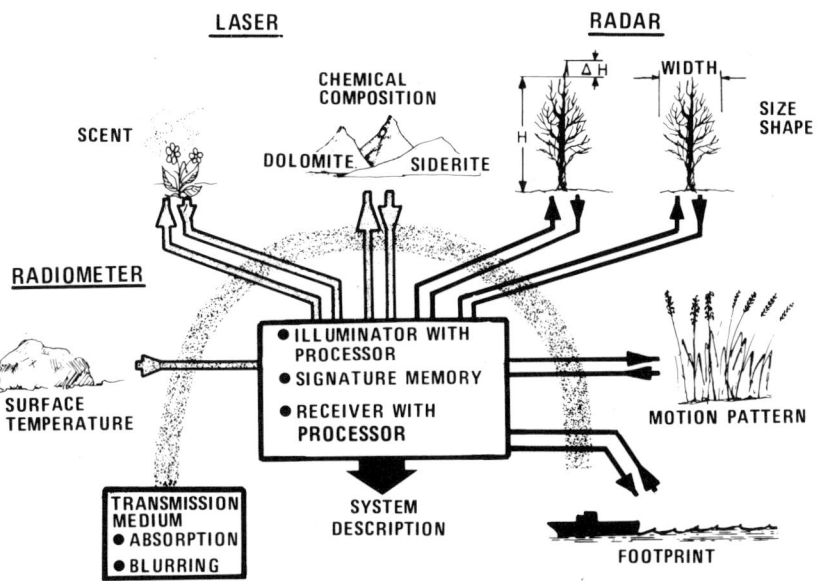

Figure 1. An artist's conception of the multisensor principle. A phenomenon or an object is characterized in several signature domains: shape, size, motion pattern, material composition, surface temperature and footprint.

In developing this new field, applying radio science principles to the problem of detection and classification of objects or phenomena with hitherto unidentifiable properties, the need arose for a unified treatment of these principles and their applications. This book is an endeavor to satisfy this demand. It attempts to provide a coherent and brief outline of the underlying theory and of the basic principles behind present-day radar remote probing methodology. It was felt that by limiting the scope to radar remote sensing, it should be possible to present a unified picture, applying essentially the same fundamental concepts to each of the three problem areas involved: the target, the environment against which the target is viewed and the intervening propagation medium. From such a unified treatment it is possible to synthesize the optimum radar illumination (matched illumination) and also "matched filter" reception, taking into consideration the characteristics of target, background and propagation medium. What this means in essence is the following: if we know the signature of the target of interest sufficiently well, and if we know the background and the intervening transmission medium, we can put these signatures into the memory of a computer and instruct the computer to adapt the radar illumination and the processing of the received signal to the target in an optimum manner.

The book is intended for advanced students or scientists with good all-around knowledge of electromagnetics and inversion theory, but it does not assume previous acquaintance with this specialty. Simple first-principle, first-order mathematics have been used to produce a book that can be read from cover to cover without supporting literature. If read in this way, it is the author's hope that the book will reveal some of the main ideas of the adaptive radar concept applied to challenging and relevant problems of today. Furthermore, it is hoped that the reader, like the author, will find the current trend within the general field of radio science a stimulating one.

Dag T. Gjessing

ACKNOWLEDGMENTS

Looking back on the work in preparing this book, I feel a deep sense of gratitude to a number of people: colleagues, assisting staff and employers. To no one, however, do I owe a more sincere thanks than to my secretary, Eva Roedsrud, and my associates Jens Hjelmstad and Terje Lund.

Only through their expert assistance, conscientious cooperation and loyal support was it possible to complete this book in the time available. Questions helped to remove obscurities, discussions gave stimulus and added momentum, and the companionship contributed to satisfaction and animation.

Among colleagues at the Norwegian Defense Research Establishment, my affiliation for more than 20 years, I am particularly indebted to Karl Holberg. His wisdom and his creative and responsive mind have been a continual source of inspiration and encouragement.

Likewise, I wish especially to thank my colleagues at the Royal Norwegian Council for Scientific and Industrial Research, Hans C. Christensen and Robert Major, for stimulation and animation and for loyal support at crucial moments.

I also wish to express my appreciation to Marit Ekstrand, who tolerantly and very ably handled the arduous task of preparing all of the art work.

Finally, I wish to thank my family: my wife, Toril, my daughter, Randi, and my son, Trygve. The effort involved in preparing this book, during evening hours, has also required their loyal support.

Dag T. Gjessing heads the Environmental Surveillance Technology Program, Royal Norwegian Council for Scientific and Industrial Research, Kjeller, Norway. He received his undergraduate degrees from Bergen University and London University (Electrical Engineering), and his graduate degree in Geophysics from Oslo University. Dr. Gjessing was previously chief scientist at the Norwegian Defense Research Establishment, Kjeller, Norway, and has also done research for Stanford University. He is a member of various international scientific organizations, among them the Norwegian Academy of Technical Sciences, the Norwegian Geophysical Society, IEEE, and he is president of the International Scientific Radio Union, Commission F on Remote Sensing and Propagation. He has 100 papers, of which more than 50 have appeared in international scientific journals and books on such topics as radiogeophysics, meteorology, measuring techniques (remote probing), antennae and communication systems. Dr. Gjessing is the author of a 1978 Ann Arbor Science publication, *Remote Surveillance by Electromagnetic Waves for Air–Water–Land.*

CONTENTS

1. Introduction 1

2. Rough Surface Scattering: A Summary of
   Basic Theory 5

   Correlation Properties of Electromagnetic Waves
      Having Different Frequency (Bandwidth
      Considerations) 8
   Correlation Properties of Scattered Electromagnetic
      Field in Space: Angular Distribution 18
   Temporal Correlation Properties of a Scattered
      Wave (Motion Pattern Considerations, Doppler) . 22

3. Signature of General Targets in Relation to a Multifrequency Adaptive Radar System 27

   Spatial Signature of an Object (Wavenumber
      Matching) 29
   Motion Pattern Analysis (Temporal Signature) 30

4. Signature of the Sea Surface as a Target Background:
   Background-Adaptive Radar Concept 35

   Spatial Signature (Wavenumber Spectrum of
      Sea Waves) 35
   Azimuthal Distribution of Radio Waves
      Scattered from the Sea Surface, Propagation
      Direction of Ocean Wave 41

Temporal (Doppler) Signature of the Sea Surface.... 45
Mutual Coherence of Radio Waves Scattered from
    the Ocean Surface and Ships, Space/Time
    Coherence............................... 48

5. **Fundamentals of Radio-Wave Propagation through the Atmosphere: Propagation Medium-Adaptive Radar** .. 53

    Line-of-Sight Propagation 54
    Propagation Mechanisms Involving Scattering and
        Diffraction 64
    Basic Relationships in Over-the-Horizon Scatter
        Propagation............................. 65
    Calculation of Pulse Distortion in Terms of
        Radiometeorological Parameters............. 69
    Calculation of Bandwidth 71
    Correlation Distance of Field Strength 74
    Antenna Gain Degradation 76
    Wavelength Dependence of Scattered Power...... 79
    Radiometeorological Parameters n and a in
        Relation to Routine Meteorological
        Observations............................ 81
    Scattering from Atmospheric Layers (Waves) 85
    Scattering by Particles (Rainfall)................103
    Diffraction of Radio Waves by Obstacles.........104
    Basic Theory of Diffraction....................106
    Absorption Phenomena.......................126
    Absorption by Gases........................126

6. **"Intelligent" Radar: Adaptive Radar Systems in Relation to Target, Terrestrial Background and Propagation Medium**......................................133

References..145

Index..151

CHAPTER 1

INTRODUCTION

Progress in radio science during the last decade has been remarkable. At first, this progress was stimulated primarily by interest in radio communications and by military needs. A period followed in which space research was a driving force. Now progress is much a result of an upsurge of interest in problems concerned with the environmental sciences, earth's resources, pollution and conservation.

In parallel with this general progress in radio science, we are witnessing a development toward a new generation of powerful technology, a new generation of devices and components. Solid-state microwave sources can efficiently be controlled by microprocessors to give a desired field configuration. Dynamic matched receiver filtering is provided by devices based on surface acoustic waves. New optoacoustic systems together with charge-coupled devices hold great promise in regard to intricate on-line processing and pattern recognition.

The radio scientist is facing a challenging and inspiring future, one of matching new technological achievements to important applications in the field of remote detection and identification.

This book is dedicated to the following concept: most of the existing detection/identification systems do not make optimum use of all of the a priori information on the object of interest of which one generally is in possession. By knowing something about the geometric shape and motion pattern of an object on

2 ADAPTIVE RADAR

which our attention is focused, an optimum illumination and detection system which adapts itself to this target against a terrestrial background can be designed. Figure 1.1 is an artist's conception of the general detection/identification problem based on a priori information about the target.

If one has general information about the background (additive noise) and specific information about the target of interest, a radar system can be designed which adapts itself to the target to give a maximum overall signal-to-noise ratio. This means, in essence, that we are faced with the consideration of three filter functions. One is constituted by the transmission medium between the observation platform and the target. (This intervening medium gives rise to multiplicative noise or distortion.)

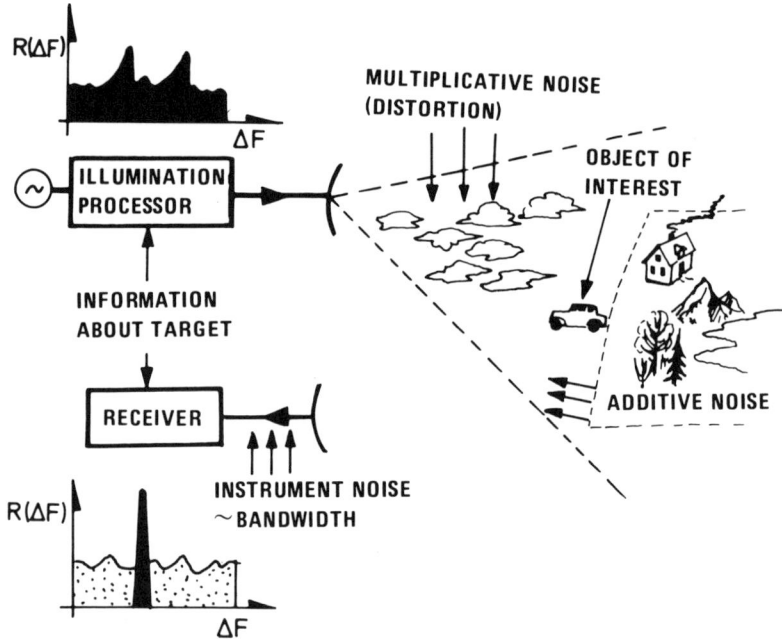

Figure 1.1 Symbolic representation of the general detection/identification problem. If we have a priori information about the object of interest, the intervening propagation medium and the background, we can tailor the illumination waveform to optimize the detection/identification capability.

The second filter we shall have to consider is determined by the terrestrial background against which the target is viewed. Third, we shall consider the target itself. The more detailed information we require about the particular target per unit time, the more wide-banded must our radar illuminator be, and the larger must the bandwidth be of the propagation medium between the scene to be investigated and the observation platform. We must, therefore, tailor the illuminating waveform to obtain maximum information about the object of interest and at the same time ensure minimum adverse influence from the intervening transmission medium (See Figure 1.1).

To provide "matched illumination" in relation to a target, we shall have to structure the illumination both in space and in time. If we have at our disposal a radar system which can be amplitude-modulated (a pulsed radar), we shall have to shape the radar pulse to obtain maximum influence on the returned radar pulse by the particular target of interest. If we have at our disposal a radar illuminator which can be structured in the frequency domain, then we should compose an illuminating frequency spectrum to obtain constructive interference by all the reflecting facets of the target.

Although in principle, there is little difference between the two approaches, in this book we shall concentrate on the multifrequency radar system. As we shall see, this system lends itself directly to simple computer control in a manner which is very familiar to the computer scientist.

Having structured the illumination in the time domain for optimum coupling to the target, it remains to shape the phasefront in space so as to obtain maximum coupling to the particular reflecting structure of interest. Reference 1 shows that by making use of a matrix antenna (two-dimensional broadside array) as the radar receiver, the phase and amplitude at each receiver element can be controlled by a computer system to provide an antenna system that is matched to the phasefront of the wave system, which is reflected back from the target of interest. Simultaneously, the waves originating from the terrestrial background are suppressed. Such adaptive phased-array systems have been the subject of many publications [1-3]. In this book we shall limit ourselves to multifrequency radar. It should be noted, however, that phenomenologically, and also in regard to the mathematical treatment, there is little difference between a spaced antenna system and one which makes use of many coherent frequencies.

CHAPTER 2

ROUGH SURFACE SCATTERING: A SUMMARY OF BASIC THEORY

Consider a volume dv within which permittivity ϵ and/or field strength $\vec{E}_0$ vary. For a plane wave incident on the scattering volume, we have:

$$\vec{E}_0 = \vec{E}_1 e^{j(\omega t - \vec{k}_1 \cdot \vec{x})}$$

where $\vec{k}_1$, as shown in Figure 2.1, is the wave number of the incident wave.

If $\vec{k}_s$ is the wave number of the scattered wave and θ is the scattering angle (angle between $\vec{k}_1$ and $\vec{k}_s$), we can define a difference wave vector:

$$\vec{K} = \vec{k}_1 - \vec{k}_s$$

such that

$$|\vec{K}| = \frac{4\pi}{\lambda} \sin \theta/2$$

which is the Bragg conditions for scattering in the direction θ. λ is the wavelength of the radio wave.

6   ADAPTIVE RADAR

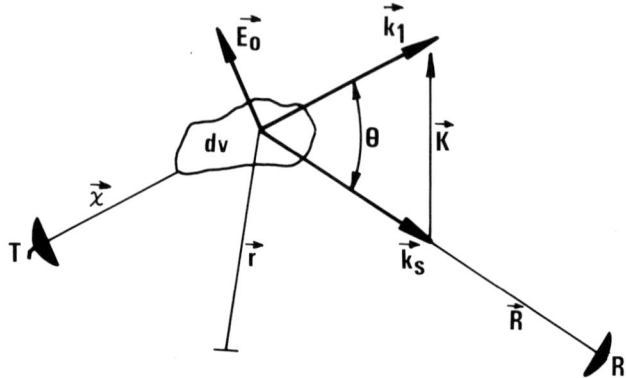

Figure 2.1 Scattering from an elementary scattering element.

If the permittivity ϵ or the field strength $\vec{E}_0$ within the elementary scattering volume dv differs from the average values of ϵ and $\vec{E}$, a dipole moment is set up:

$$d\vec{P} = \Delta\epsilon \, dv \, \vec{E}_0 = \Delta\epsilon \, dv \, \vec{E}_1 \, e^{j(\omega t - \vec{k}_1 \cdot \vec{x})}$$

At the distance $\vec{R}$ from this dipole moment, we will have a polarization potential $\vec{\Pi}$, given by:

$$d\vec{\Pi} = \frac{\Delta\epsilon \, dv \, \vec{E}_1}{\epsilon \, 4\pi \, R} e^{j(\omega t - \vec{k}_1 \cdot \vec{x} - \vec{k}_s \cdot \vec{R})}$$

Integrating this over all contributions to the polarization potential within the total volume V illuminated by the transmitter and seen by the receiver (the scattering volume), we have:

$$\vec{\Pi} = \frac{1}{4\pi R} \int f_\epsilon(\vec{r}) \, \vec{E}_1(\vec{r}) \, e^{j(\omega t - \vec{K} \cdot \vec{r})} \, d\vec{r}$$

where $\vec{r}$ is a position vector, such that

$$\vec{k}_1 \cdot \vec{x} + \vec{k}_s \cdot \vec{R} = \vec{K} \cdot \vec{r}$$

where $\vec{K} = \vec{k}_1 - \vec{k}_s$.

Knowing the polarization potential $\vec{\Pi}$, we can calculate the field strength $\vec{E}_s$ from the well-known relationship:

$$\vec{E}_s = \nabla\nabla \cdot \vec{\Pi} + k^2 \vec{\Pi}$$

Provided the scattering volume (spatial region illuminated by the radar) is small in comparison with the distance R ($R \gg V^{1/3}$):

$$k^2 \vec{\Pi} \gg \nabla\nabla \cdot \vec{\Pi}$$

such that the scattered field $\vec{E}_s$ is given by:

$$\vec{E}_s \approx k^2 \vec{\Pi}$$

The scattered field resulting from an integral of elementary scattering elements is given by:

$$\vec{E}_s = \frac{k_s^2}{4\pi R} \int_V \vec{E}(\vec{r}) \, \epsilon(\vec{r}) \, e^{-j\vec{K}\cdot\vec{r}} d^3\vec{r} \qquad (2.1)$$

omitting the time factor $e^{j\omega t}$.

where $\vec{K} = \vec{k}_1 - \vec{k}_s$
$\vec{k}_1$ = wave number of incident field
$\vec{k}_s$ = wave number of scattered field

Thus, if θ is the scattering angle (angle between $\vec{k}_s$ and $\vec{k}_1$), we have:

$$|\vec{K}| = \frac{4\pi}{\lambda} \sin \theta/2$$

where λ = the wavelength of the electromagnetic wave

Note that Equation 2.1, which is derived from Maxwell's equations, is perfectly general and does not consider the nature of the scattering object.

The $E(\vec{r})$ and $\epsilon(\vec{r})$ functions may be stochastic, in which case statistical descriptions must be used (spatial autocorrelation functions, spatial spectra), or we may be dealing with ordered variations, in which case well-behaved analytical functions may be used [1,4-7].

CORRELATION PROPERTIES OF ELECTROMAGNETIC WAVES HAVING DIFFERENT FREQUENCY (BANDWIDTH CONSIDERATIONS)

Let us now simplify our approach and direct our attention to a one-dimensional scattering object. We combine the various factors contributing to the scattered field into one, namely one which is directly related to the scattering cross section as a function of distance.

We define the function f(z) as the *delay function*. This has dimension field strength such that the scattering cross section as a function of distance z along the direction of propagation (z is measured along the direction of $\vec{K}$) is obtained by squaring the f(z) function. From Equation 2.1, we have:

$$\vec{E}_s(\vec{K}) \sim \int f(\vec{z}) \, e^{-j\vec{K}\cdot\vec{z}} \, d\vec{z} \qquad (2.2)$$

where $K = \omega/c$

Note that the delay function f(z) tells us how the scatterers are distributed along the direction of propagation. We now want to investigate what information we can obtain about the distribution in depth of the scatterers constituting the scattering object by illuminating the scattering object with a spectrum of electromagnetic waves (Figure 2.2). Using the terminology of the communication engineer, we wish to calculate the "bandwidth" of the reflecting object. The question that we ask is: if we illuminate an object that is characterized by a delay function $f(\vec{z})$ with a set of coherent (mutually correlated) electromagnetic waves, how are the coherence properties of the scattered wave influenced by the shape of the scattering object [by the delay function $f(\vec{z})$]?

Let us start with the general expression for the scattered field (Equation 2.2)

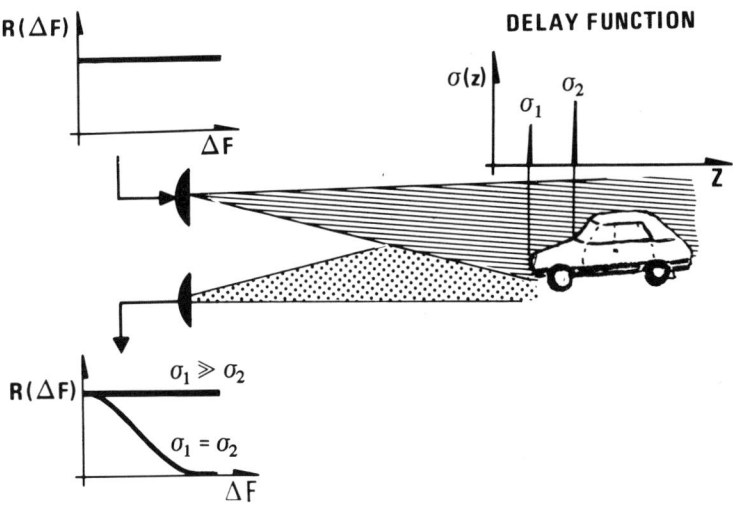

Figure 2.2 Illumination of an object with a set of correlated radio waves. The coherence properties of the scattered waves are determined by the distribution in depth f(z) of the scatterers constituting the scattering object.

10 ADAPTIVE RADAR

$$E(K) \sim V(\omega/c) \sim \int f(z) e^{-j\vec{K}\cdot\vec{z}} dz \qquad (2.3)$$

This equation states that the field strength/voltage at frequency ω (the amplitude spectrum) is the Fourier transform of the delay function $f(z)$.

A convenient way of expressing the degree of coherence (the bandwidth properties) between different waves is by the correlation $R(\Delta\omega)$ between their complex amplitudes.

$$R(\Delta\omega) = \frac{\overline{E(\omega) E^*(\omega + \Delta\omega)}}{|E(\omega)|^2}$$

The overbar denotes statistical average. The average may be a time average or an ensemble average. Hence:

$$\overline{E^*(\vec{K})E(\vec{K}+\Delta\vec{K})} \sim \int f(\vec{z})e^{j\vec{K}\cdot\vec{z}}d\vec{z} \quad \int f(\vec{z}+\vec{r})e^{-j(\vec{K}+\Delta\vec{K})(\vec{z}+\vec{r})} d\vec{r}$$

i.e.,

$$\overline{E^*(\vec{K})E(\vec{K}+\Delta\vec{K})}$$

$$\sim \iint e^{-j(\vec{K}+\Delta\vec{K})\cdot\vec{r}} d\vec{r} \; \overline{f(\vec{z}) f(\vec{z}+\vec{r})} \; e^{-j\Delta\vec{K}\cdot\vec{z}} d\vec{z} \qquad (2.4)$$

Note that such a statistical average can be obtained in several ways. We can select the product $E(\omega) E^*(\omega + \Delta\omega)$ for a given $\Delta\omega$ at n different values of ω and thus obtain an ensemble of n independent samples, provided $E(\omega_2)$ is uncorrelated with $E(\omega_2)$, etc. (see below).

Alternatively, if we are dealing with an object which is time-variable (e.g., the sea surface), we can make use of a time average.

The second factor of Equation 2.4 is recognized as the complex autocorrelation function of the delay function. The phase factor of the autocorrelation function (the term $e^{-j\vec{\Delta K}\cdot \vec{z}}$) oscillates rapidly with $\vec{\Delta K}$, since the space coordinate $\vec{z}$ (the distance from the illuminator to the scattering object) is large in comparison with the incremental range $\vec{r}$. Equation 2.4 reduces to:

$$\overline{E(\vec{K})\,E^*(\vec{K}+\vec{\Delta K})} \sim e^{-j\vec{\Delta K}\cdot \vec{z}} \int R(\vec{r})\, e^{-j(\vec{K}+\vec{\Delta K})\cdot \vec{r}}\, d\vec{r} \qquad (2.5)$$

provided z, the distance from the radar to the object, is large in comparison with the size of the object and, provided the "beat-wavelength" (see Figures 2.4 and 2.5), $2\pi/\Delta K$ is also large in comparison with the size of the object. Hence:

$$R\left(\frac{\Delta\omega}{c}\right) = \frac{\overline{E\left(\frac{\omega}{c}\right) E^*\left(\frac{\omega+\Delta\omega}{c}\right)}}{A}$$

$$= \frac{e^{-j\frac{\Delta\omega}{c}z} \int R(r)\, e^{-j\frac{1}{c}(\omega+\Delta\omega)\cdot r}\, dr}{A} \qquad (2.6)$$

where A = a normalizing factor of the form $\int R(r)\, dr$.

Equation 2.6 states that the complex correlation in the frequency domain of waves scattered back from an object characterized by a given delay function is the Fourier transform of the autocorrelation function R(r) of this delay function. By measuring the bandwidth of a reflecting object, we obtain directly statistical information about the delay function characterizing the object.

The modulus of the autocorrelation function $R(\Delta\omega/c)$ is directly the Fourier transform of the R(r) function. This is depicted in Figures 2.2 and 2.4.

12 ADAPTIVE RADAR

Let us now, to ensure a physical understanding of the general principles involved, calculate the bandwidth for a set of objects characterized by simple analytical delay functions. First, let us consider an object which can be characterized by an exponential delay function:

$$f(t) = e^{-\alpha t}$$

The 1/e width of this delay function is:

$$t_0 = \frac{1}{\alpha}$$

The voltage at frequency ω is the Fourier transform of this exponential delay function. Hence, omitting a constant:

$$V_1(\omega) = (\alpha + j\omega)^{-1}$$

Similarly, the voltage V_2 at frequency $(\omega + \Delta\omega)$ is given by:

$$V_2(\omega + \Delta\omega) = [\alpha + j(\omega + \Delta\omega)]^{-1}$$

The normalized complex autocorrelation of these voltages is then given by:

$$R(\Delta\omega) = \frac{\int_{-\infty}^{\infty} (\alpha + j\omega)^{-1} [\alpha - j(\omega + \Delta\omega)]^{-1} \, d\omega}{\int_{-\infty}^{\infty} (\alpha^2 + \omega^2)^{-1} \, d\omega} \qquad (2.7)$$

Solving this integral, we derive the following expression for the modulus of the autocorrelation function:

$$R(\Delta\omega) = \sqrt{1 + \left(\frac{\Delta\omega}{2\alpha}\right)} \qquad (2.8)$$

To conform with the nomenclature of the communications engineer, let us, on the basis of Equation 2.3, calculate the power spectrum and through this the "bandwidth" in the conventional manner.

Inserting $\Delta\omega = 0$ into Equation 2.6, we obtain the following expression for the "power spectrum":

$$W(\omega/c) = \overline{E(\omega/c) \, E^*(\omega/c)} \sim \int R(r) \, e^{-j\frac{\omega}{c} r} \, dr \quad (2.9)$$

where, as before, $R(r)$ is the autocorrelation of the delay function $f(z)$.

Table 2.1 gives the power spectrum and bandwidth of a set of objects having delay functions which are simple analytical functions.

Figure 2.3 shows the "radar signatures" in a normalized form of some targets, all of size 100 m. Note the marked influence of target shape. Note also that if the target has a periodic structure, such as an exponentially damped sinusoidal variation, the correlation function in the frequency domain peaks up at a frequency separation ΔF, which is different from zero and determined by the period δz:

$$\Delta F = \frac{c}{2\delta z}$$

Note also that the width of the "bandwidth function" is determined by the degree to which the sinusoidal target is damped (truncated). From simple convolution considerations we know that if a sinusoidal function is exponentially damped such that Δz is the $1/e$ width of the damping function, the width of the $R(\Delta F)$ function corresponding to $R(\Delta F) = 1/2$ is given by:

$$\Delta F_{\frac{1}{2}} = 0.16 \frac{c}{\Delta z}$$

14 ADAPTIVE RADAR

Table 2.1 The Bandwidth Properties (Multifrequency Radar Signature) of Some Object Classes Expressed Analytically

Distribution in Depth of Scatterers	Frequency Dependence of Reflecting Surface	Half Power Bandwidth
Exponential	$W(\omega) = \dfrac{\sigma_o^2}{\left(\dfrac{c}{z_o}\right)^2 + \omega^2}$	$\Delta F_{1/2} = 0.16 \dfrac{c}{z_o}$ Hz
Gaussian	$W(\omega) = \dfrac{\pi}{4} \dfrac{\sigma_o^2}{\left(\dfrac{c}{z_o}\right)^2} e^{-\dfrac{\omega^2 z_o^2}{8 c^2}}$	$\Delta F_{1/2} = 0.37 \dfrac{c}{z_o}$ Hz
Rectangular	$W(\omega) = 4\sigma_o^2 \left(\dfrac{\sin \dfrac{z_o}{2c}\omega}{\omega}\right)^2$	$\Delta F_{1/2} = 0.44 \dfrac{c}{z_o}$ Hz
Triangular	$W(\omega) = \dfrac{64 \sigma_o^2}{\left(\dfrac{z_o}{c}\right)^2} \left(\dfrac{\sin \dfrac{z_o}{4c}\omega}{\omega}\right)^4$	$\Delta F_{1/2} = 0.64 \dfrac{c}{z_o}$ Hz
Discrete Scattering Centers	$W(\omega) = 4\sigma_o^2 \left(\cos \dfrac{z_o}{2c}\omega\right)^2$	$\Delta F_{1/2} = 0.25 \dfrac{c}{z_o}$ Hz

ROUGH SURFACE SCATTERING 15

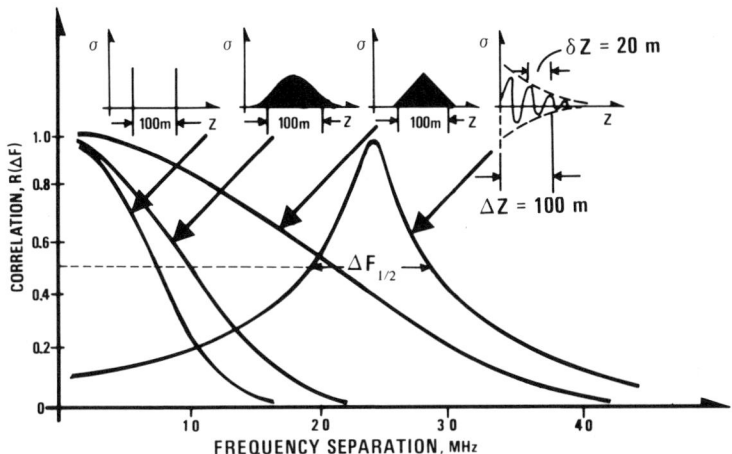

Figure 2.3 Knowing the geometric shape of an object (spatial distribution of scatterers), we can calculate the radar signature. If the target manifests itself as a limited number of discrete scattering centers, the radar signature will be the sum of cosinus relationships.

If, therefore, the scattering object is the sea surface with wavelength δz (see Chapter 5), and if the radar illuminator gives an exponential intensity distribution over the spot-size Δz, the relative wavelength resolution of our wave radar is given by:

$$\frac{\Delta F_{1/2}}{\Delta F_o} = \frac{0.16 \frac{c}{\Delta z}}{\frac{c}{2\delta z}} = 0.32 \frac{\delta z}{\Delta z} \qquad (2.10)$$

Hence, if, for example, we illuminate a sea-surface area containing 10 ocean wavelengths, our wavelength resolution is 3%.

To ensure a thorough mathematical and physical understanding of the factors involved, we shall dwell for a moment on Equation 2.9, Table 2.1 and Figure 2.3.

16 ADAPTIVE RADAR

We have seen that the autocorrelation function in the frequency domain of the signal scattered back from a rough surface (rough in terms of carrier frequency wavelength) is the Fourier transform of the autocorrelation function of the delay function, whereas the amplitude spectrum of the scattered field is the Fourier transform of the delay function itself. From general Fourier analysis [8] and Equation 2.6, we know that if we Fourier-transform a nonsymmetrical function, a complex correlation function results. Thus, if the object is at distance z_o, as illustrated in Figure 2.4, the delay function will be nonsymmetrical and the autocorrelation function R(ΔF) will oscillate with period $c/2z_o$. The envelope of the correlation function R(ΔF) is obtained by taking the modulus of the complex autocorrelation function as in Equation 2.6 above.

Thus, by measuring the complex autocorrelation function in the frequency domain, obtained by using a multifrequency radar system, we obtain information about the distance to the

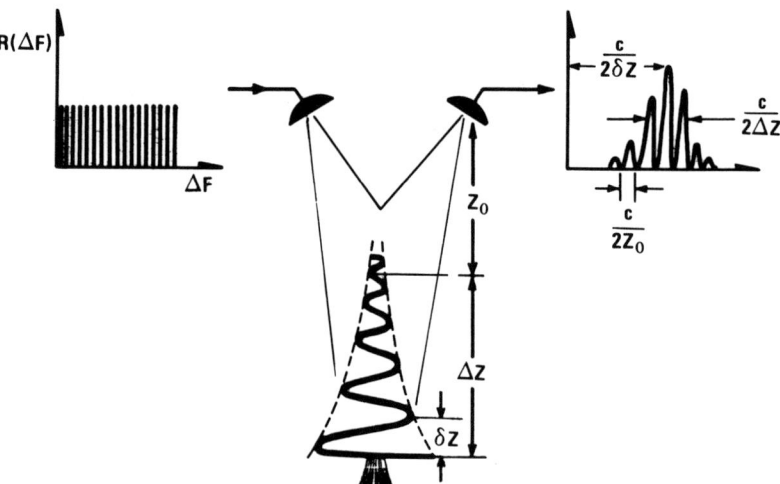

Figure 2.4 Illumination of an object with a set of electromagnetic waves with different frequencies. The correlation function in the frequency domain of the reflected wave gives information about the distance to and size and shape of the object.

object, the size of the object and about its shape. Physically this mathematical statement can be visualized from Figure 2.5 [9].

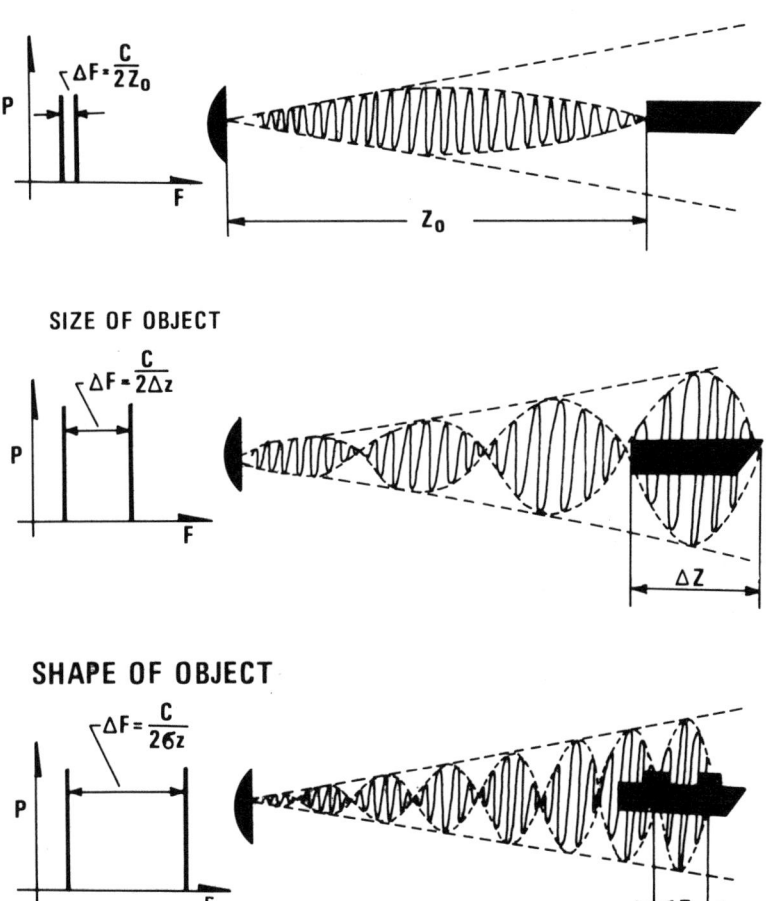

Figure 2.5 The distance z_o to the object is determined by transmitting two microwave frequencies with spacing $c/2z_o$. The size of the object is obtained in the same manner by transmitting two frequencies with spacing $c/2\Delta z$. The shape is obtained by transmitting waves with the larger frequency separation $c/2\delta z$ [9].

To measure the distance to the object (assumed large in comparison with the size of the object) we transmit two microwave frequencies and adjust their frequency separation to obtain one spatial beat period between transmitter and target. In practice, what this means is the following. Assume that one is interested in knowing when an approaching target passes the 100-km distance zone. Disregarding for the moment Doppler phenomena (discussed later), we shall transmit two radar frequencies with mutual separation $\Delta F = c/2z_o = 1.5$ kHz. If, for example, the length of this target at 100 km range should be 100 m long, to be of interest, we should check if the two transmitted frequencies with mutual spacing $\Delta F = c/2\Delta z$ where $\Delta z = 100$ m are in phase when scattered back from the target. This calls for two frequencies with mutual spacing 0.48 MHz. Note that the carrier frequency only enters into the question because it influences the scattering cross section of the target.

CORRELATION PROPERTIES OF SCATTERED ELECTROMAGNETIC FIELD IN SPACE: ANGULAR DISTRIBUTION

We have completed the section on the correlation properties of electromagnetic waves having different frequency. We have seen that by measuring the degree to which waves having different frequency are correlated, we obtain information about the longitudinal distribution of the scatterers. If we are dealing with a thin reflector (zero distribution in depths), the bandwidth of the reflector is very large. Conversely, if the scatterers are distributed over a large region in space, the bandwidth is small.

We shall now focus the attention on the transverse distribution of the scattering elements constituting the scattering object. To reveal this transverse structure, we shall use another characteristic property of electromagnetic waves, namely, spatial correlation properties. We shall illuminate the scattering object with a single frequency, and at the receiving site we shall make use of a set of antenna elements distributed along a baseline or in a plane perpendicular to the line joining the receiving array and the scattering object. At each element, we shall measure the amplitude and the phase of the impinging wave. On the basis of these point-observations of field strength, information about the target can be extracted. This will be the subject of this section.

First, consider Figure 2.6, illustrating the scattering process. We illuminate a surface with a single wave. Depending on the properties of the surface, we obtain a certain field-strength distribution of the scattered field $\vec{E}_s(\vec{r})$.

We shall now calculate the angular distribution $P(\theta)$ of the waves scattered from this surface. From Equation 2.1, we find the following expression for the scattered field in terms of the field-strength distribution over the scattering surface:

$$\vec{E}_s(\vec{K}) \sim \int \vec{E}(\vec{r}) e^{-j\vec{K}\cdot\vec{r}} d\vec{r} \qquad (2.11)$$

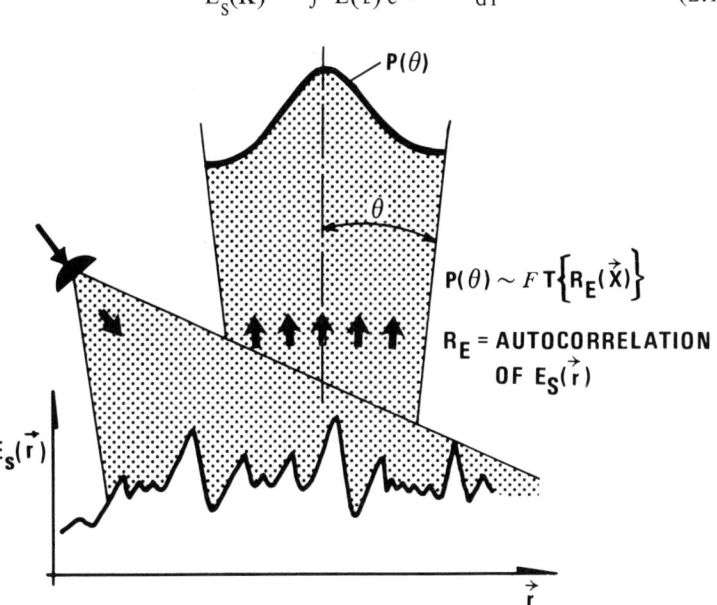

Figure 2.6 The geometry of the backscattering process. The illuminating field gives rise to a scattering field-strength distribution $E_s(\vec{r})$.

20 ADAPTIVE RADAR

Since

$$K = \frac{4\pi}{\lambda} \sin \theta/2$$

this equation tells us how the scattered field is distributed in direction θ. From this we shall derive the angular power distribution $P(K)$ as follows:

$$P(\vec{K}) \sim E_s(\vec{K}) E_s^*(\vec{K})$$

$$\sim \int\int E(\vec{x}) E(\vec{x}+\vec{r}) e^{j\vec{K}\cdot\vec{x}} e^{-j\vec{K}(\vec{x}+\vec{r})} d^3\vec{x}\, d^3\vec{r}$$

$$\sim \int d^3\vec{r}\, e^{-j\vec{K}\cdot\vec{r}} \int d^3\vec{x} E(\vec{x}) E(\vec{x}+\vec{r}) \qquad (2.12)$$

The second integral is immediately recognized as the spatial autocorrelation $R_E(\vec{r})$ of the field-strength distribution.

Hence

$$P(\vec{K}) \sim P(\sin \theta) \sim \int R_E(\vec{r}) e^{-j\vec{K}\cdot\vec{r}} d^3\vec{r} \qquad (2.13)$$

This equation tells us that the angular spectrum (radiation pattern) of the scattered wave is the Fourier transform of the field-strength distribution over the scattering region when this distribution is expressed statistically in terms of its spatial autocorrelation.

This is a relationship which is very well known from antenna theory: the radiation pattern (angular power distribution) of an antenna with aperture A is obtained by the Fourier transform of the field-strength distribution over this aperture. Thus, if our $E(\vec{x})$ function is a rectangular one, implying that the field strength is evenly distributed over the antenna aperture, the angular power distribution is of the form $(\sin \theta)/\theta$, and the beamwidth $\beta = \lambda/D$, where D is the aperture size. Conversely, if the field-strength distribution over the aperture is of the (sin x)/x form, the angular distribution of the scattered wave is a rectangular one.

We shall be using these simple relationships extensively in the subsequent sections. Now let us return to Equation 2.13 and use this as the basis for studying another important property of the scattered field, namely, the spatial correlation of field-strength.

In the discussion which we have just completed, we considered the case of a "transmitting antenna." Now let us consider the case of a receiving one. Our receiving antenna consists of a set of antenna array elements which permits us to measure amplitude and phase at each array element. The power reaching this array antenna is distributed as $P(\theta)$ over an angular region. Applying the inverse Fourier transform of Equation 2.13 above, it is intuitively obvious that we obtain information about the spatial correlation properties of the field-strength:

$$R_E(\vec{r}) \sim \int P(\vec{K}) e^{j\vec{K} \cdot \vec{r}} d\vec{K} \qquad (2.14)$$

This equation tells us that the spatial correlation of the scattered field is the Fourier transform of the angular power distribution.

We see from Figure 2.7 that if we are to resolve an object of transverse extent Δx by a receiving antenna array at distance R from the object, we shall have to measure the field strength distribution over a spatial region $L_x = R\lambda/\Delta x$.

Summing up these findings, we should note that by measuring the field-strength distribution across a broadside array (amplitude and phase at each array-point), we obtain direct information about the transverse scattering properties of the scattering object.

It takes little imagination to see the analogy between the spatial autocorrelation function and the autocorrelation function in the frequency domain discussed above. In the multifrequency case, we can "filter out" certain longitudinal spatial distributions of the scattering object by using frequency filters. In the

22 ADAPTIVE RADAR

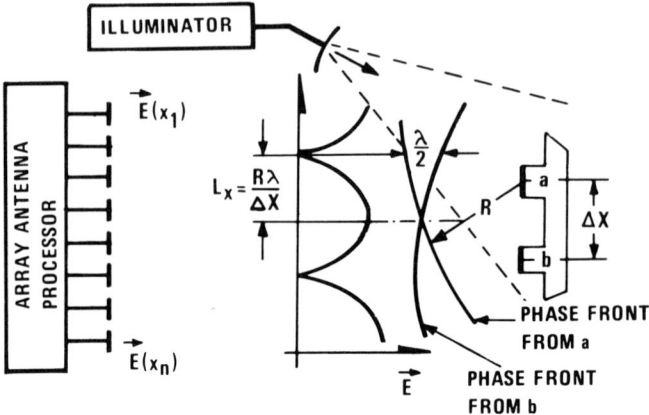

Figure 2.7 Transverse distribution of field strength. The spatial autocorrelation of field strength is the Fourier transform of the angular power spectrum. The size (width) Δx of the object manifests itself as a "transverse interferogram."

case of the broadside array, we can "filter out" certain transverse properties of the scattering elements by providing a "spatial filter." This means that we can make an adaptive system by adjusting the amplitude and phase of the receiving antenna elements to match the wavefront of the wave system which is reflected back from the target, while the waves originating from the terrestrial background against which the target is viewed can be suppressed. Such adaptive phased-array systems have been reported [2,3]. In this chapter we shall limit ourselves to referring to Figure 2.7, where a symbolic presentation of the phased array concepts is given. We shall also point out the very striking analogy between the multifrequency adaptive system and that involving phased arrays.

TEMPORAL CORRELATION PROPERTIES OF A SCATTERED WAVE (MOTION PATTERN CONSIDERATIONS, DOPPLER)

We have shown that if a rigid body is illuminated by a set of electromagnetic waves having different frequencies, information about the distribution in depth (along the direction of wave propagation) is obtained by studying the correlation properties of the scattered waves. Specifically, if two electromagnetic waves with frequency separation ΔF are illuminating the object, we obtain information about a particular irregularity scale L (spatial spectrum component $K = 2\pi/L$) where L is related to frequency separation ΔF, as follows:

$$L = \frac{c}{2\Delta F}$$

Let us now assume that a rigid body characterized by the delay function $f(z)$ (distribution in depth of the scattering elements) is moving with velocity $\vec{V}$. We now want information about the Doppler shift to which this frequency ΔF is subjected.

We base this calculation on the basic Doppler equation

$$f = \frac{1}{2\pi} \vec{K} \cdot \vec{V} \qquad (2.15)$$

Here $K = 2\pi/L$, where L is the scale-size to which the frequency pair ΔF is matched. The Doppler shift associated with the difference frequency ΔF is therefore:

$$f = \frac{1}{2\pi} \vec{K} \cdot \vec{V} = \frac{1}{2\pi} \frac{2\pi}{L} V \cos \phi$$

$$= \frac{2\Delta F}{c} V \cos \phi \qquad (2.16)$$

where ϕ is the angle between the wave vector $\vec{K}$ and the velocity vector $\vec{V}$.

Hence, if we illuminate a moving object with two electromagnetic waves with frequency spacing ΔF (coupled to scale-size $L = c/2\Delta F$), this frequency ΔF is subjected to a Doppler shift f which is proportional to ΔF and to the velocity V of the object. The power associated with the Doppler frequency f is the same as that associated with the frequency ΔF and expressed by Equation 2.9.

Thus, knowing the shape and velocity of the target (see Figure 2.3), the Doppler spectrum can be calculated on the basis of Equations 2.9 and 2.16.

Let us consider a flexible object. The spectrum of irregularity scales constituting the scattering body are moving at different velocities. Consider one particular scale L which is distributed throughout the scattering body. Let us assume that the width of the velocity distribution of the scattering elements characterized by the scale L (Fourier component $K = 2\pi/L$) is δV. This velocity distribution will, obviously, give rise to a Doppler spectrum the width of which is determined by the velocity spread:

$$\Delta f = \frac{1}{2\pi} \vec{K} \cdot \vec{\delta V} \qquad (2.17)$$

In terms of the illuminating frequency ΔF, we get the following expression for Doppler broadening:

$$\Delta f = \frac{2\Delta F}{c} \delta V \cos \phi \qquad (2.18)$$

Figure 3.5 shows the Doppler shift to which each radio frequency separation ΔF is subjected, when the scattering object is a rigid ship characterized by two scattering centers 100 m apart moving at a velocity of 20 knots (37 km/hr).

Now let us finally consider the case where the scatterers are distributed over the entire area illuminated by the radar beam. If the beamwidth is β, the direction of the backscattered radio wave will vary between $-\beta/2$ and $+\beta/2$, such that the direction of the wave vector $\vec{K}$ will vary over the angle β.

As seen from the basic Doppler equation above, this variation in $\vec{K}$ will lead to a Doppler spread in excess of that caused by the velocity spread δV.

Let us assume that the scatterers are filling the radar beamwidth and that they are moving at speed V_c along a direction which is normal to the center line of the radar beam. Scatterers located at the center line obviously give rise to no Doppler shift, whereas scattering elements located at extreme positions give a Doppler shift:

$$\Delta f = \pm \frac{2\Delta F\, V_c}{c} \sin \beta/2$$

as seen from Equation 2.16. For narrow antenna beams, therefore, the Doppler broadening caused by a cross-beam drift velocity V_c of the scattering elements is given by [10]:

$$\Delta f = \frac{2\Delta F}{c} V_c \cdot \beta \qquad (2.19)$$

This Doppler broadening effect is in practice of little importance when dealing with targets of finite size, but it may be significant when dealing with a strong cross-beam ocean surface current (see Chapter 4).

CHAPTER 3

SIGNATURE OF GENERAL TARGETS IN RELATION TO A MULTIFREQUENCY ADAPTIVE RADAR SYSTEM

The discussion above, dealing with established physical principles, forms the basis for a target-adaptive radar system. Figure 3.1 shows a specific target-adaptive radar system that is under development by the author's institution for the purpose of studying ocean waves and ship and aircraft signatures. A microprocessor controls a set of frequency generators, to

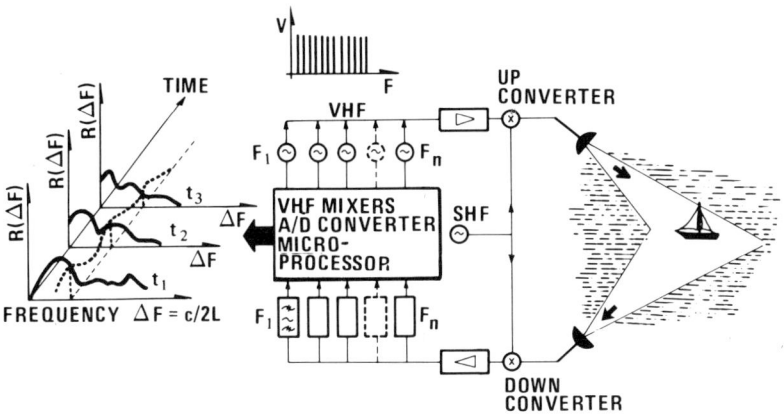

Figure 3.1 Schematic diagram of a target-adaptive radar system being developed in its simplest form by the author's institution. A microprocessor system measures distance to object, size and shape, in turn. By keeping track of the time history of the relevant descriptive parameters (z_o, Δz, δz), the motion pattern (speed, pitch, yaw, roll) is obtained.

illuminate the object of interest with a suitable set of frequencies. The receiver consists of a set of narrow-band filters capable of receiving each of the transmitted frequencies with the appropriate Doppler broadening. The signals from these receivers are being analog-to-digital converted in the same microprocessor, which performs the appropriate computations and displays the autocorrelation functions in the frequency domain (distribution in depth of the scattering object) within each time frame of interest. Thus, if the object is at rest and if the transmission medium is frozen, one would expect to have a set of identical correlation functions along the time axis. If, however, the object changes, for example, aspect angle (yaw, roll and pitch motion of a ship) we would also see a structure in the time domain. From Figure 2.5, it is self-evident that the microprocessor can be programmed to investigate each of the target properties sketched in Figure 2.5: range, and size and shape of object. Referring now to the receiving end of Figure 3.1, we see that if we know the position and the size and shape of the object at rest, the filters F_1 and F_n can be made infinitesimally narrow, and we obtain a system with extreme sensitivity.

In general, if we know the distribution in depth $\sigma(z)$ of the object of interest (i.e., the delay function of the object), we should illuminate the object with a set of frequencies distributed as some inverse function of the delay function characterizing the object [1]. The result of this illumination would ideally be a reflected wave whose correlation properties in the frequency domain are reduced to a delta function requiring an infinitesimally narrow bandwidth for detection. This is, not surprisingly, in accord with the classical concepts from information theory: it requires zero bandwidth to convey a known message; it takes zero bandwidth to detect a sinusoidally varying signal when its period is known and observation time is unlimited.

The task we are now facing is to form the basis for an optimum estimation of target parameters to provide the computer system with a protocol, based on which a strategy for a general adaption procedure can be developed.

SIGNATURE OF TARGETS 29

We shall confine ourselves to a discussion of ship targets against an ocean background. Furthermore, we shall characterize the ship target against the sea background in three signature domains: space, motion pattern and space/time coherence.

SPATIAL SIGNATURE OF AN OBJECT (WAVENUMBER MATCHING)

In the preceding chapters we have shown, on the basis of general physical principles, that the correlation properties of radio waves scattered back from an object at distance z_o are given directly as the Fourier transform of the delay function characterizing the object. Thus, if the object is a rectangular one, with length Δz at a distance z_o, the envelope of the autocorrelation function in the frequency domain of the returned radio waves is simply a $(\sin x)/x$ relationship, the width of which is $c/2\Delta z$ and under which envelope there are sinusoidal oscillations, with period $c/2z_o$.

We shall now consider the spatial signatures of ships. To calculate a detailed signature as it appears at the radar station, we must know how the scattering cross section is distributed along the target. In the practical case, this will not be a smooth function. From experience we know that the "bulk" scattering cross section is made up of a finite number of discrete scattering centers which physically manifest themselves as corner reflectors. The delay function of a ship target (the Fourier transform of which is the radar signature) will, then, probably consist of a limited number of unequally spaced δ functions. From the conventional diagrams of scattering cross section, plotted against the azimuthal aspect angle, we can, to some degree, deduce the delay functions. Figure 3.2 shows a set of target delay functions together with their resulting radar signatures [1,11,12].

Curve A of Figure 3.2 shows the multifrequency-radar signature of a 100-m ship characterized by an even distribution of scattering centers, whereas curve B shows the signature of one consisting of two dominating scatterers 100 m apart.

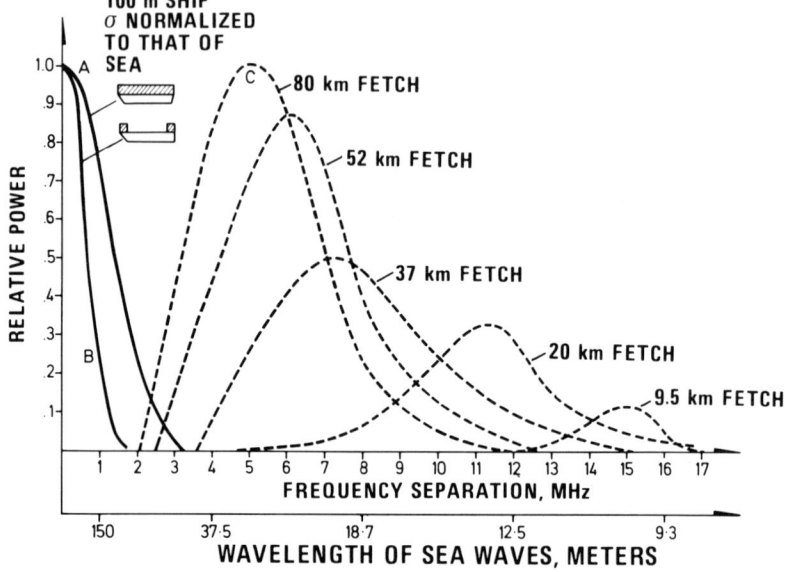

Figure 3.2 Signatures of ships compared with sea-surface signatures. The sea calculations are based on the JONSWAP experiments, and the relative power scale is arbitrary.

In contrast, Figure 3.3 [13] shows the relative scattering cross section of different classes of ship targets. These curves show that the scattering cross section of ship targets varies substantially from class to class. Note, for instance, that the scattering cross section corresponding to curve A (small ship observed from an elevation of 470 ft), is something like 60 dB smaller that that of a destroyer observed from the same platform.

MOTION PATTERN ANALYSIS (TEMPORAL SIGNATURE)

In the previous chapter we saw that, knowing the delay function of the target, an illumination function can be structured (composition of wavelengths) to obtain maximum signal-to-

SIGNATURE OF TARGETS 31

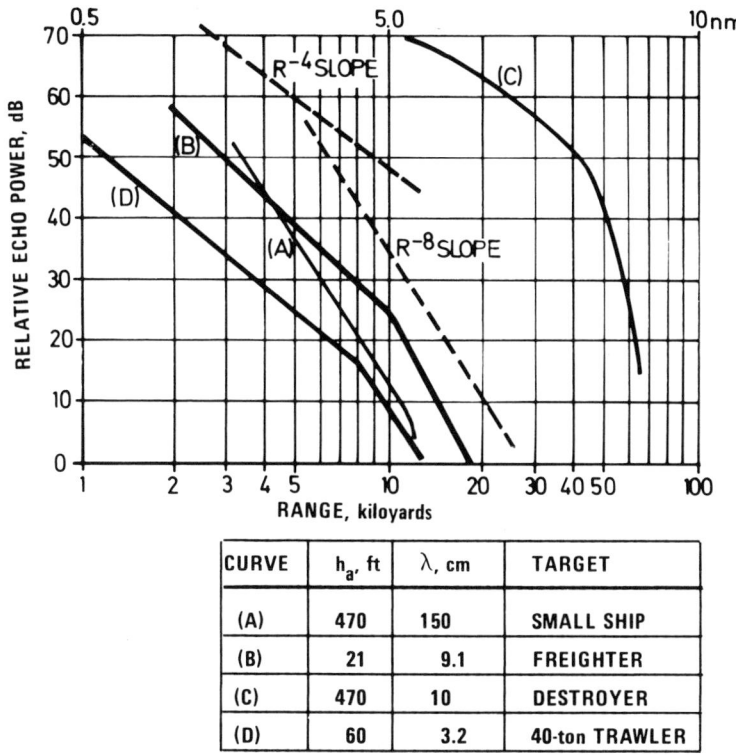

CURVE	h_a, ft	λ, cm	TARGET
(A)	470	150	SMALL SHIP
(B)	21	9.1	FREIGHTER
(C)	470	10	DESTROYER
(D)	60	3.2	40-ton TRAWLER

Figure 3.3 Relative scattering cross section of various ship targets. Three different observation platforms were employed at heights of 470, 60 and 21 ft above sea level (after Nathanson [13]).

noise ratio for the target of interest, and also to obtain a target identification capability. We shall now introduce another aspect, namely that of temporal (Doppler) filtering. We illuminate the target and its background with a set of electromagnetic waves having different wavelengths, and we observe the Doppler shift and Doppler broadening associated with each of the backscattered radar waves. For the purpose of specificity, let us base the following discussion on Figure 3.1, showing a particular experimental multifrequency radar system.

In this system the radar transmitter consists of six mutually correlated frequency generators. The receiver, on the other hand, has six very narrow-banded receiving amplifiers, which are synchronized to the transmitters. The frequencies of the 6 transmitters and the 6 receivers are chosen to provide 15 different frequency separations. We now measure the Doppler shift and broadening associated with each of these 15 frequency pairs simultaneously (see Chapter 2). This means that we can look at 15 different scale sizes, both in the target and at the sea surface, and we can measure the rate at which these scales move one-half-beat frequency wavelength (one-half wavelength along the difference frequency scale).

Note that if the number of independent frequency lines available are organized in this manner, we obtain only one $E(\omega) E^*(\omega+\Delta\omega)$ product per frequency separation $\Delta\omega$. We therefore have to make use of time-averaging to obtain estimates which are statistically significant.

Introducing the concept "motion pattern filtering," the reader is referred to Figure 3.4. Here we have shown a target in the form of a ship and have indicated two dominant target scales. One scale is constituted by the bow and the stern of the ship (length of ship), the other by the ship's funnels. We choose two frequency pairs (three different but mutually correlated radio frequencies) to match one pair to the scale associated with the length of the ship, and another frequency pair to the scale-size associated with the spacing of the funnels. Note that since the ship is rigid and in translatory motion, both these scale-sizes are moving at the same velocity V_T.

As shown above, this ship velocity gives rise to a Doppler shift given by:

$$f = \frac{2\Delta F}{c} V_T \cos \phi \qquad (3.1)$$

where ϕ is the angle between the direction of the ship and that of the radar beam.

SIGNATURE OF TARGETS   33

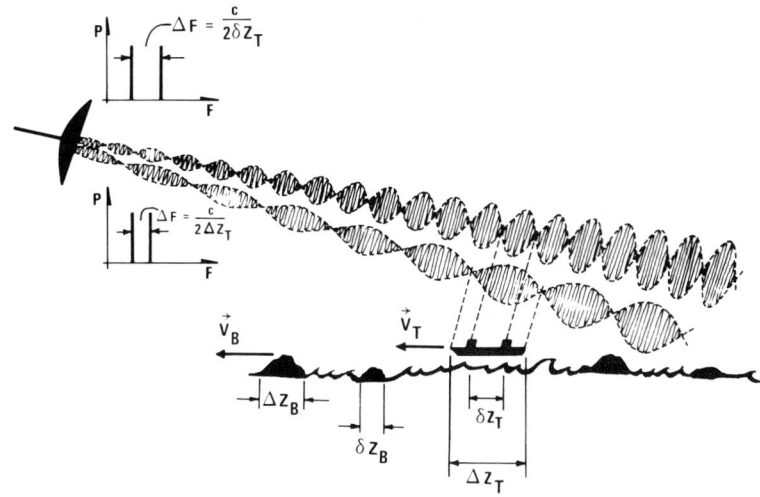

Figure 3.4 Simple three-frequency adaptive radar system. One frequency pair is coupled to the total length of the ship target; the other pair is coupled to another set of scattering centers. The echo signal from the target competes with that from the sea surface only where the irregularity scales of the ocean are the same as those of the target.

Figure 3.5, curve A, shows the Doppler distribution for a 50-m ship characterized by two dominating scattering centers moving at a radial speed of 20 knots (37 km/hr). This Doppler distribution is obtained directly from the P vs ΔF relationship in Figure 3.2 by linear scaling in accordance with Equation 3.1 above. Curve B of Figure 3.5 shows the Doppler signature of a 100-m ship of speed 20 knots (37 km/hr).

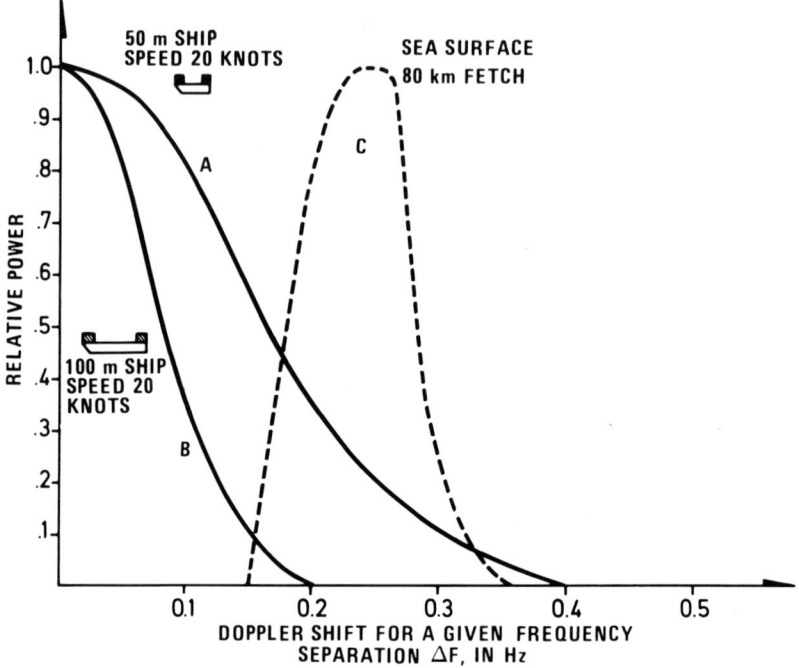

Figure 3.5 The Doppler spectrum of the dispersive sea waves compared with the Doppler shift of the "matched radar illumination" caused by the motion of a ship. The sea spectrum is based on the JONSWAP experiments. The relative power scale is arbitrary.

CHAPTER 4

SIGNATURE OF THE SEA SURFACE AS A TARGET BACKGROUND: BACKGROUND-ADAPTIVE RADAR CONCEPT

SPATIAL SIGNATURE (WAVENUMBER SPECTRUM OF SEA WAVES)

In the preceding chapters we have shown that if we know how the scattering capability of a scattering surface varies along the direction of radio wave propagation, then we can calculate the "bandwidth" of this scattering surface. We characterized the surface by a delay function $f(z)$, and we saw that the bandwidth function (the autocorrelation function in the frequency domain of the scattered field) is obtained by Fourier-transforming the spatial autocorrelation of this delay function $f(z)$. From the mathematical formulations we could draw the following conclusions: if we multiply the field $E(K)$ at wave number K with the complex conjugate $E^*(K + \Delta K)$, and apply appropriate averaging schemes, we obtain information about the scattering cross section at the scale $L = \pi/\Delta K$. We also saw that if the scattering elements of linear extent L were moving, a Doppler shift of the radio wave scattered back by the ensemble of these spatially distributed scattering elements resulted.

We shall now apply these general theoretical results to the sea surface. Specifically, we shall discuss a method by which we can determine experimentally the statistics of the delay function. Note that a priori this has nothing to do with the physical height distribution of the sea waves, although intuitively we would expect the spatial spectrum of scattering intensity to bear some resemblance to the spatial spectrum of the physical wave-height [14-24].

This statement can be amplified as follows: consider first a single sinusoidal sea wave (wavelength L_w) which has a rough surface. By rough we mean an irregular surface structure where the irregularity scale is comparable with that of the radar carrier wave, the wave number of which is $\vec{K}$. Assume first that the intensity of the irregularities at scales comparable with $\vec{K}$ is constant over many sea wavelengths. Under such simplified conditions we can calculate the delay function when we know the wave height and the wavelength, since there is a simple empirical relationship between the incidence angle of the radio wave and the corresponding scattering coefficient (the scattering coefficient increases approximately exponentially with the incidence angle in the interval from gracing incidence to a few degrees [25]). Under these conditions we would get maximum return from the points on the sea surface where the slope is a maximum (second derivative is zero). Hence, we would obtain a simple periodic delay function with period equal to that of the sea wave. Unfortunately, however, the real situation is very much more complex. Dynamic shear forces in the boundary layer give rise to variations in the local stability factor. As a result of these, the fine scale structure riding on the sea waves (capillary waves) is damped in a periodic fashion along the sea wave [26,27]. In addition, there are shadowing effects giving rise to diffraction phenomena [7], and for certain geometrical configurations we may experience local focusing [23], complicating the situation still further. In conclusion, therefore, because of factors such as:

- local tilting,
- shear force damping,
- shadowing effects, and
- focusing effect

it is not an easy task to relate scattering cross section to wave height through an analytical approach, even when we consider one simple well-behaved sea wave. If we consider a realistic situation, the problem is even more complicated.

The signature of the sea surface as well as that of the target is normalized. To calculate the contribution from the sea surface to the scattered signal, relative to that from the ship, we shall have to know the total scattering cross section of the ship, and also that of the sea surface. Much information has been collected on the bulk scattering cross section from ships and sea surfaces under varying degrees of sea state. Figure 4.1 [25] shows the scattering cross section of the sea surface as a function of depression angle for various sea states. Note that σ_o is the ratio of the actual sea-surface scattering cross section relative to that of the illuminated sea surface. Thus, if the spot size of the illuminating beam is A m^2, for a depression angle of some 3° and a sea index (sea state) of 5, the actual scattering cross section of the sea surface is −30 dB relative to the illuminated area.

In broad terms, there are three rather distinctly different mechanisms influencing the spatial surface structure of the sea. Mechanical disturbances set up by the force of a wind field generate a spectrum of sinuosoidal waves which are controlled by gravitational forces. These waves are known as gravity waves, and their motion is given by the very simple dispersion relationship for deep water, namely $\omega = (g\,K)^{1/2}$, the wave number $K = 2\pi/L$ and the frequency $\omega = 2\pi/T$, where T is the period of the water wave and L its wavelength. The complete expression for the gravity wave, angular frequency including the case of shallow water is given by $\omega = (g\,K)^{1/2} \tanh^{1/2}(KD)$ where D is the water depth. These are the simplest types of ocean disturbances [28].

If, however, these waves are propagating in an environment where for several reasons a velocity shear develops (e.g., on account of the presence of a stationary interphase, such as a beach), an instability may develop (e.g., Calvin-Helmholtz instabilities) causing the simple wave structure to form breaking

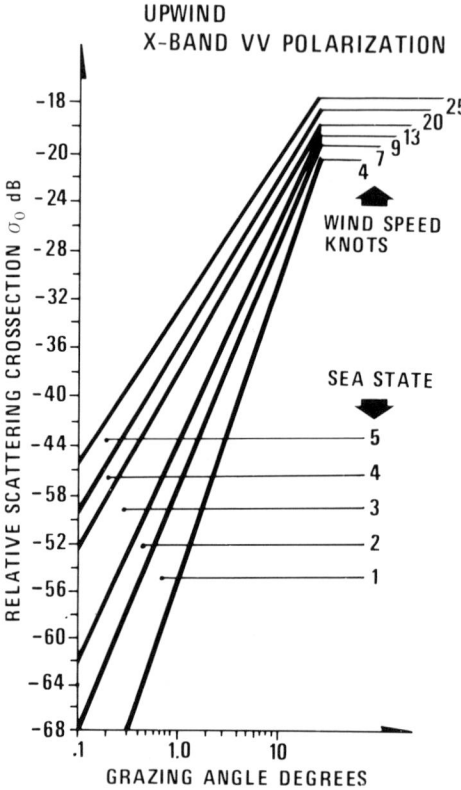

Figure 4.1 Scattering cross section of sea surface for an X-band radar as a function of depression angle and various sea states and wind velocities (after Sittrop [25]).

waves, which again leads to a turbulent and stochastic situation [26,27].

Finally, the boundary layer wind field (the long fetch action of which gives rise to gravity waves and turbulence) also influences the water surface directly. This "footprint of turbulent air motion" complicates the surface structure of the sea still further [29].

The result of all these dynamic mechanisms influencing the surface structure of the sea is a wide spectrum of surface irregularities ranging from gravity waves with hundreds of meters wavelength down to capillary waves in the centimeter region. In this book, however, we shall confine ourselves to

SIGNATURE OF SEA SURFACE 39

the wave-number region of ocean irregularities carrying most of the energy, namely, the region dominated by gravity waves, and, for the reasons listed above, we shall not be able to give anything but qualitative results. Experimental investigations are, however, being carried out at several research organizations such as the U.S. Naval Research Lab, DFVLR in Germany and the author's institution. Based on these investigations, empirical relationships are about to become available.

Figure 4.2 shows the general principle for the interaction of electromagnetic waves with the sea surface. The microwaves couple directly to the capillary waves that are superimposed

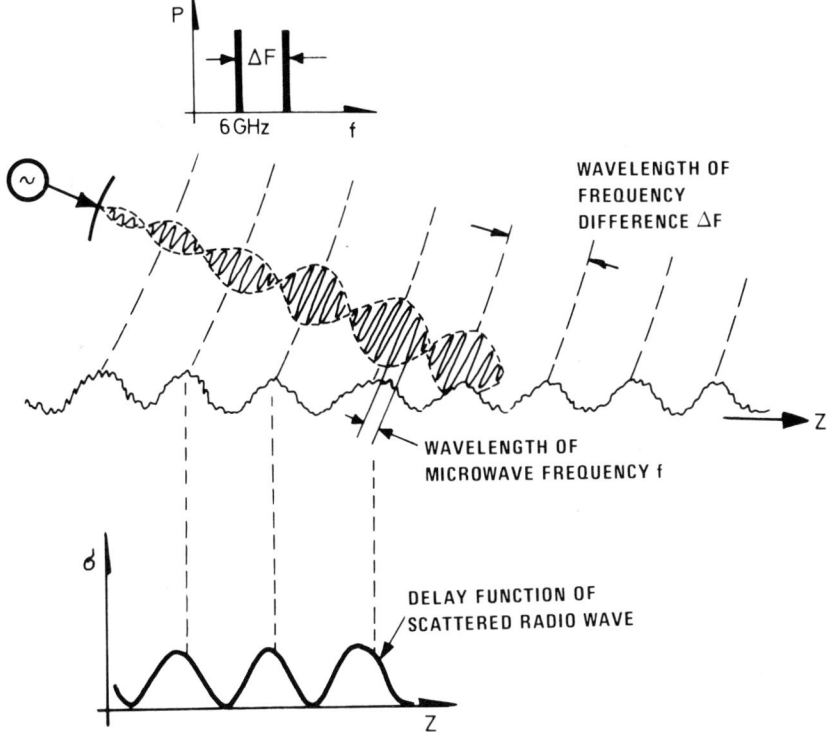

Figure 4.2 Interaction of electromagnetic waves with the sea surface. Microwaves couple directly to capillary waves, which are superimposed on longer-wavelength phenomena (gravity waves). The envelope of the beat frequency pattern couples to the gravity wave pattern.

40 ADAPTIVE RADAR

on longer wavelength phenomena (gravity waves). The envelope of the beat frequency pattern couples to the gravity wave.

The following calculations will be based on the JONSWAP experimental results reported by Hasselman et al. [30].

Knowing the ocean wave-height spectra in the conventional time domain, it remains, before we can calculate the radar signature, to convert the temporal spectra to spatial spectra applying the dispersion relationship for deep water gravity waves:

$$\omega = (g|\vec{K}|)^{1/2} \tag{4.1}$$

where the wave number $\vec{K}$ is two-dimensional such that the modulus becomes:

$$|\vec{K}| = (K_x^2 + K_y^2)^{1/2} \tag{4.2}$$

Since $\omega = 2\pi f$, where f is the temporal frequency of the wave, and since the spatial frequency (wave number) is $|\vec{K}| = 2\pi/L$, where L is the wavelength, the relationship between temporal frequency and wavelength becomes:

$$L = \frac{g}{2\pi f^2} \tag{4.3}$$

Thus, from the JONSWAP experimental results, we can calculate the wave height vs wavelength spectrum.

It only remains to select a set of radio waves to match their beat frequency to the ocean waves (see Figure 4.2). Noting as before that this matching beat frequency is given by:

$$\Delta F = \frac{c}{2L}$$

we can synthesize the "radar signature" of the ocean wave as in the case of the target. The results are shown as curve C in Figure 3.2. Please note that the vertical scale of this is arbitrary. Referring to this figure, the following should be noted:

1. Since the ocean waves exhibit a spatial periodicity, the sea-surface signature will have its maximum for a frequency separation which is different from zero. For a fetch of, for example, 52 km, the sea-surface signature peaks up for a frequency separation of approximately 6 MHz. The ship signature, however, peaks up at zero frequency since the target is assumed to be nonperiodic.
2. The signature of the sea surface, as well as that of the target, is normalized. To calculate the contribution from the sea surface to the scattered signal relative to that from the ship, we shall have to know the total scattering cross section of the ship, and also that of the sea surface. Much information has been collected on the bulk scattering cross section from ships and on the sea surface under varying degrees of sea state [25]. Similarly, the total "bulk" scattering cross section of different classes of ship targets has been measured [13].

In conclusion, referring to Figure 3.2, note that, in the frequency range up to 2 MHz, there is practically no influence of the sea-surface irregularities. Above this frequency, the sea surface dominates. From the point of view of "filtering out" the target from a sea background, this is an ideal situation.

AZIMUTHAL DISTRIBUTION OF RADIO WAVES SCATTERED FROM THE SEA SURFACE, PROPAGATION DIRECTION OF OCEAN WAVE

In the previous chapter we established a simple relationship between the difference frequency ΔF and the irregularity scale L (wavelength of sea wave) to which this beat frequency ΔF would couple. By simple Bragg scatter considerations we saw that if the sea surface is illuminated by two electromagnetic waves with difference frequency ΔF, it is irregularity scales

42 ADAPTIVE RADAR

(ocean wavelength) of length $L = c/2\Delta F$ that are responsible for the backscattering.

We shall now calculate the azimuthal distribution of the radio wave scattered from the sea surface. For this purpose we shall consider the situation where one single plane ocean wave (wavelength L_w) is illuminated by a matched frequency pair ($\Delta F = c/2L_w$). Let us assume that the antenna aperture is d and the distance from the antenna to the illuminated spot on the sea surface is R such that the width of this spot is D meters. (D = $(\lambda/d)R$, where λ is the radio wavelength). Referring now to Figure 4.3, illustrating the geometry, we see that if the radar antenna is pointed in a direction such that the phasefront of the ocean waves coincides with that of the radio wave, constructive interference will take place. Turning now the antenna through an azimuthal angle β, a situation arises where the radio waves which are scattered from one half of the illuminated

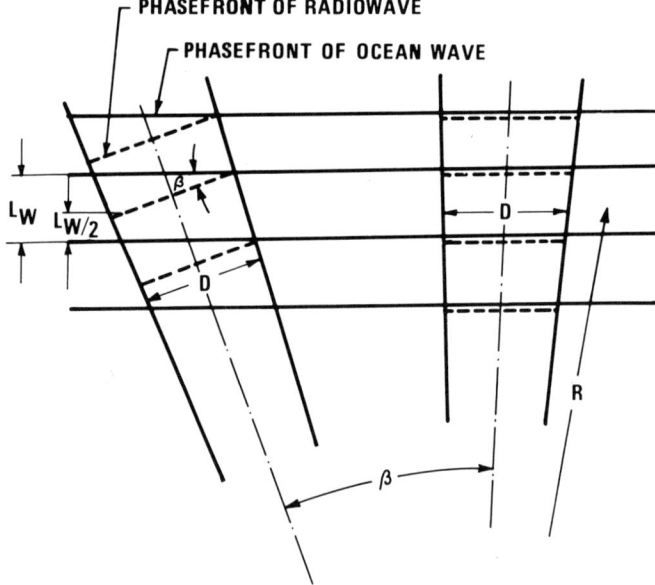

Figure 4.3 Geometry of constructive and destructive coupling of electromagnetic waves to a plane ocean wave.

ocean area cancel those scattered from the other half by destructive interference. This leads to a null in the angular power spectrum of the backscattered waves. As seen from Figure 4.3 the azimuth angle giving rise to zero backscatter is given by:

$$\beta = \tan^{-1}(L/2D)$$

In terms of antenna aperture d, range R and radio wavelength λ, the beamwidth of the backscattered wave is given by

$$2\beta = 2\tan^{-1}(L\,d/2\lambda R) \qquad (4.4)$$

Then let us consider the incoherent case. Let us assume that there are turbulent irregularities superimposed on the well-behaved plane gravity waves. Let us assume that the scale ℓ_o of these irregularities as measured along the crest of the coherent gravity wave is small in comparison with the width D of the illuminated spot. Such an irregularity pattern superimposed on, and partly linked to, the gravity waves would conceivably arise if a strong local wind left a "turbulent footprint" behind on the ordered gravity waves. Breaking waves resulting from local instability phenomena would also give rise to such an irregular sea surface. Note that since these irregularities are influenced by the local wind field, they will have a velocity different to that of the coherent gravity waves. If we, therefore, were to measure the Doppler spectrum of the sea-return (see below) these turbulent irregularities would appear as "sidebands" on the Doppler component produced by the much more intense coherent gravity waves. The results of preliminary experiments [31] give support to this hypothesis.

Referring to Figure 4.4, a frequency difference ΔF, which for $\beta = 0$ couples to an ocean wave irregularity of wavelength L_1, will for the angle β couple to a wavelength L_2 which is the projection of L_1 and given by:

$$L_2 = L_1 \cos \beta \qquad (4.5)$$

44 ADAPTIVE RADAR

If the scattering intensity associated with scale L_2 is larger than that of L_1, we would experience an initial increase in scattered power with azimuth angle β. If the converse is the case, we would experience a decrease. Whether we are dealing with turbulence or gravity waves, the wave-number spectra exhibit a very marked maximum (e.g., as indicated in Figure 3.2). Note, however, that these intensity spectra are generally distinctly asymmetrical. The rate at which the intensity decreases with increasing wave number on the high wave-number side of the spectrum maximum is lower than the rate at which the intensity increases on the low wave-number side of the maximum.

The result of these spectral properties is a rather complex angular distribution of the scattered radio wave. Whereas the coherent Doppler component stemming from a plane gravity wave will have its maximum for a direction which coincides with that of the gravity wave, this is not the case for the "incoherent sidebands" of the Doppler spectrum (see below).

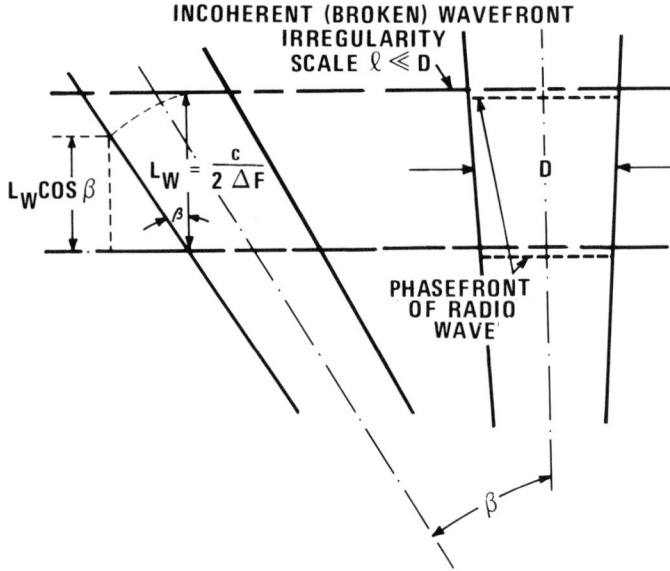

Figure 4.4 Scattering from an incoherent (broken) wavefront caused by a local wind field and/or breaking waves.

TEMPORAL (DOPPLER) SIGNATURE OF THE SEA SURFACE

In Chapter 3 we showed that frequency components with mutual frequency spacing ΔF couple to irregularity scale sizes $\Delta L = c/2\Delta F$. However, if we select a set of frequencies to match the target, the frequencies will also be matched to the corresponding irregularities of the sea surface (see Figure 3.4). We shall now consider the Doppler signature of these sea-surface irregularities.

We have already observed that the irregularity structure (the delay function) of the sea surface is determined by several mechanisms. Gravity waves are well behaved, and propagate at a velocity:

$$V = (gL/2\pi)^{1/2} \qquad (4.6)$$

where L is the wavelength.

Superimposed on these, there are "incoherent" irregularities, many of which are also of length L. These move in a random disordered fashion since they are caused by breaking gravity waves and air turbulence. They may manifest themselves as a "patchiness" in the capillary wave structure, and can be visualized as "footprints" of the air turbulence field. Their incoherent random motion pattern will, as we shall see in a moment, give rise to a broadening of the Doppler line caused by the ordered motion of the gravity wave.

Now, let us consider the Doppler signature of the gravity wave. The Doppler shift is, as we have seen in Chapter 3, given by:

$$f = \frac{1}{2\pi} \vec{K} \cdot \vec{V}$$

and, since $K = 2\pi/L$:

$$f = \frac{2\Delta F}{c} V \cos \phi$$

46 ADAPTIVE RADAR

where ϕ is the angle between the wave vector $\vec{K}$ and the velocity vector $\vec{V}$ of the ocean wave.

Hence, the Doppler shift of the frequency difference ΔF caused by the sea wave is

$$f = \frac{2\Delta F}{c} (gL/2\pi)^{1/2} \qquad (4.7)$$

There is, however, as pointed out above, a direct relationship between the ocean wavelength L and the difference frequency ΔF which couples constructively to this ocean wave, namely $\Delta F = c/2L$. Hence, the Doppler shift to which the difference frequency ΔF is subjected is given by:

$$f = (g\Delta F/\pi c)^{1/2} \qquad (4.8)$$

This dispersive relationship is shown in Figure 4.5. Basing our calculations on Equation 4.8 and on the familiar JONSWAP spectrum for ocean waves referred to above, we can calculate the Doppler shift and spectral intensity associated with each radio frequency pair subjected to the assumption that we know the relationship between scattering cross section and wave height (see above). This is shown in Figure 3.5.

Figure 4.6 ties this up with the multifrequency radar concept. We select a ΔF_1 (which is the same as selecting an irregularity scale L_1), and we measure how the frequency covariance function:

$$R(\Delta F_1, t) = V(F,t) V^*((F + \Delta F_1),t)$$

varies with time t. We then compute the power spectrum of the frequency covariance function $R(\Delta F_1,t)$. This means, as depicted in Figure 4.6, that we apply a Fourier transformation of the $R(\Delta F_1,t)$ function with respect to time t to obtain the Doppler spectrum.

SIGNATURE OF SEA SURFACE 47

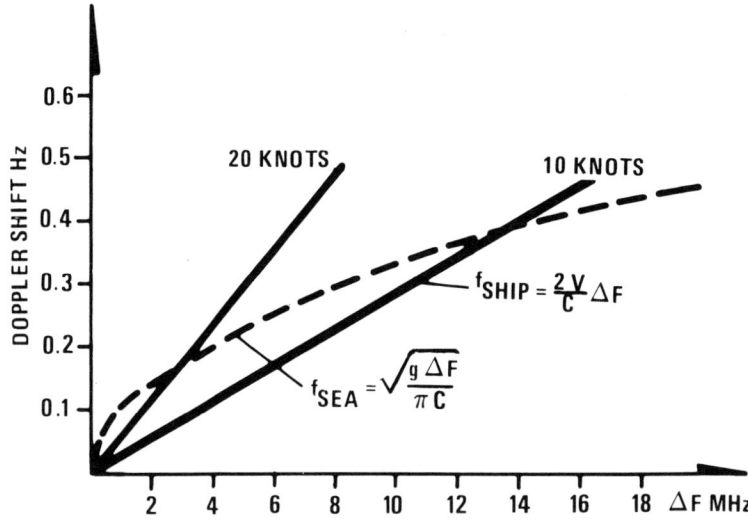

Figure 4.5 Deep-sea gravity waves are dispersive. There is, therefore, a square-root relationship between ocean wavelength (frequency difference of illuminating waves) and the Doppler shift, which is caused by the motion of the gravity wave. A rigid target gives a linear relationship between Doppler and ΔF.

If the ordered gravity waves dominate over the incoherent velocity components which ride on the gravity waves, we would expect the Doppler spectrum to have a maximum at the frequency corresponding to the dispersion relation of gravity waves (Equation 4.7), and we would expect a Doppler broadening determined by the velocity spread δV. Hence the Doppler broadening Δf is given by:

$$\Delta f = \frac{2\Delta F}{c} \delta V \cos \phi \qquad (4.9)$$

Finally, it remains to compare this Doppler spectrum of the sea surface with the corresponding one for the target (see Figure 3.5).

48   ADAPTIVE RADAR

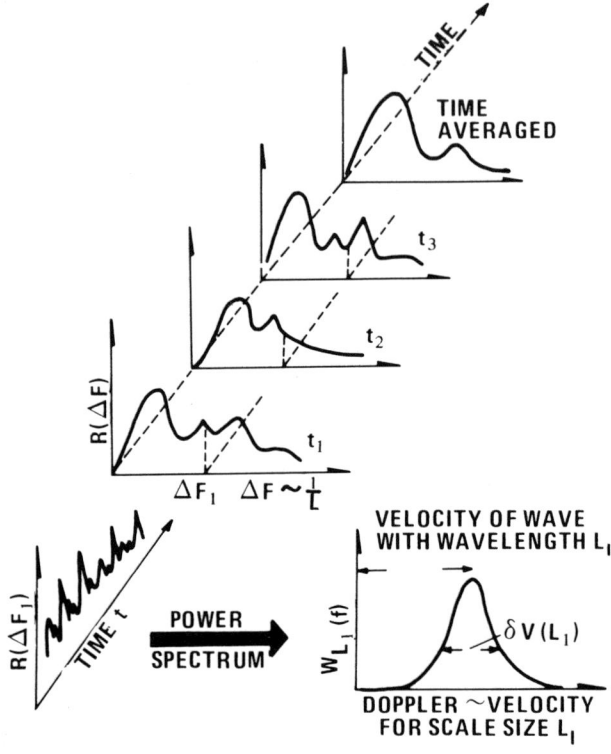

Figure 4.6 By studying the temporal variation of the frequency covariance function $R(\Delta F)$ for a given frequency ΔF_1 (scale size $L_1 = c/2\Delta F_1$), we can derive information about the motion pattern (Doppler) of the irregularity scale L_1.

Since, as we have already noted, the target is rigid such that all irregularity scales move with the same velocity V, the Doppler shift of the ship is:

$$f_{SHIP} = \frac{\Delta F}{c} \cdot 2V \qquad (4.10)$$

MUTUAL COHERENCE OF RADIO WAVES SCATTERED FROM THE OCEAN SURFACE AND SHIPS, SPACE/TIME COHERENCE

We are still addressing ourselves to the problem of "filtering out" the signature of a target in the form of a ship against a background of sea clutter. We have seen that since the sea surface exhibits a dynamic, semiperiodic irregularity structure, distributed over a large spatial region, and since the ship target manifests itself as a rigid scattering body with finite size and a nonperiodic irregularity structure, the "spatial radar signature" of this target falls into a domain which only partly overlaps that of the sea surface. Furthermore, we have established that due to the dispersive properties of ocean waves, it is also possible by Doppler filtering techniques to distinguish between ship targets and contributions from the ocean background.

There is, however, still another property of the ocean surface which is yet to be explored, that of coherence. Illuminating a rigid body with a set of electromagnetic waves in the fashion described above, each of these radio waves is subjected to a Doppler shift which is proportional to its frequency and to the velocity of the scale size to which this wave is matched. Thus, since all of the irregularity scales, constituting the total scattering cross section of the ship target for translatory motion, move with the same velocity, the Doppler frequency shifts of the various backscattered waves will be related directly to one another. This means that if we observe the Doppler shift of the frequency pair which is coupled to, say, a 50-m irregularity scale of the ship, this will be directly correlated with the Doppler shift of the frequency pair which is coupled to the 100-m scale of the target, and the relative frequency shift will be a factor of 2. This is because the scattering centers that are 100 m apart are rigidly connected to the scattering centers that are 50 m apart. Whatever changes are imposed on the velocity of the 100-m scale, the 50-m scale will follow. Referring now to Figure 4.7, this means that if we multiply the Doppler frequency resulting from the frequency pair which is coupled to the 50-m scale size by a factor of 2 and correlate this Doppler spectrum (by forming the covariance function) with that resulting from the frequency pair which is coupled to the 100-m scale size, we would expect each frequency component in the two Doppler spectra to be correlated. This is depicted in Figure 4.7, which shows the coherence function.

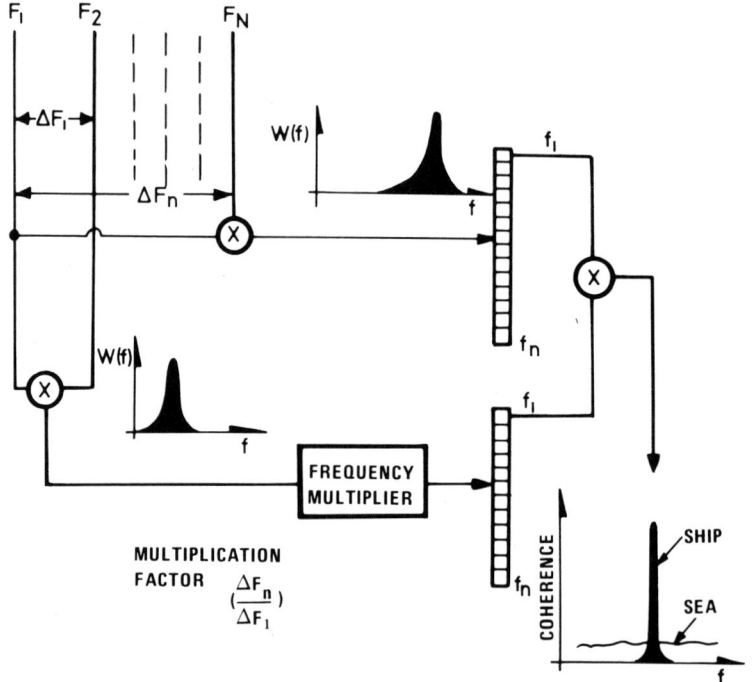

Figure 4.7 By correlating the normalized Doppler spectrum frequency components obtained at one frequency pair with that obtained at another, thus forming the mutual coherence function, the detection and identification potential of our multifrequency radar system is considerably enhanced.

Now let us consider what happens to the two frequency pairs (chosen, for example, to couple to a scale size of 50 and 100 m, respectively) when the four waves are reflected back from the sea surface. We noted in the chapter above that the 100-m wave causes a Doppler shift which is $(2)^{1/2}$ times greater than that of the 50-m wave. Furthermore, the two wave systems are not an integral part of a rigidly moving system. The 50-m wave exhibits a motion pattern which is not correlated in time or space with that of the 100-m wave.

Thus, by multiplying the Doppler spectrum resulting from the frequency pair coupled to the 50-m wave by 2, and correlating this Doppler spectrum with the one obtained from the fre-

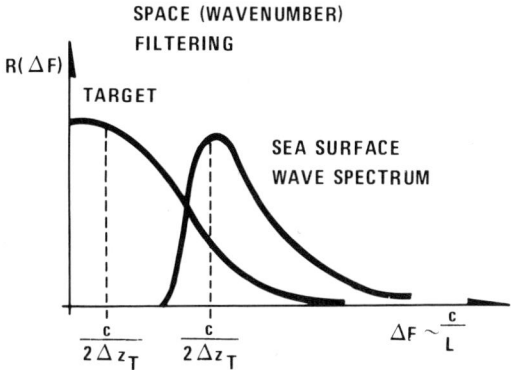
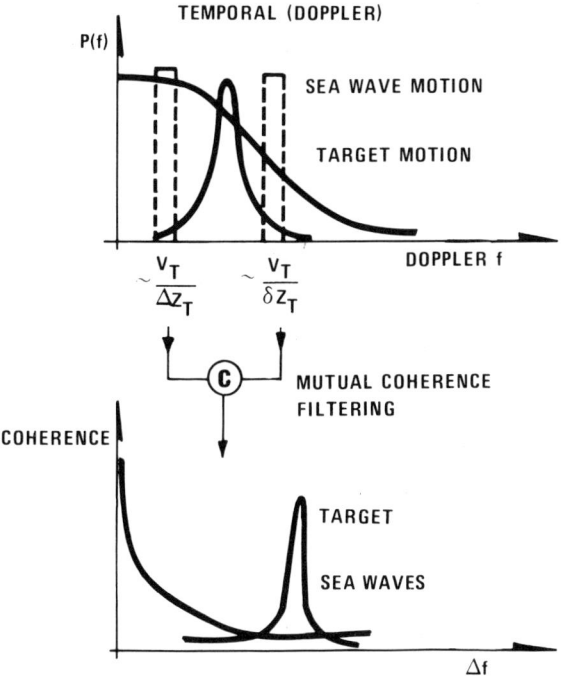

Figure 4.8 By using three sets of filters (space, temporal and mutual coherence), the detection probability and identification capability are substantially enhanced.

quency pair coupled to the 100-m wave by forming the cross-spectrum coherence function, we would not expect any significant correlation. This property of ocean waves, in contrast to the corresponding properties of ship echoes, is illustrated in Figure 4.7.

Note that to optimize the effect of these coherence filtering processes, one should make use of the same number of coherence filters as there are dominating scattering centers of the target of interest. This means that in the case of the present multifrequency radar system where we are dealing with 15 frequency pairs, we could use as many as 14 "coherence filters."

In summing up this section dealing with detection and identification of targets in an ocean environment, the reader is referred to Figure 4.8. Making use of the statistical properties of the target and background in three signature domains (space, motion pattern and space/time coherence), our detection and identification capability is substantially enhanced. The challenge is to accumulate sufficient information about the target of interest so as to adapt the radar illumination to this target signature, and at the same time optimize the signal/interference ratio on the basis of information about the sea surface.

CHAPTER 5

FUNDAMENTALS OF RADIO-WAVE PROPAGATION THROUGH THE ATMOSPHERE: PROPAGATION MEDIUM–ADAPTIVE RADAR

From the preceding chapters we have seen that to resolve the fine-scale structure of a target, we must illuminate it with broad-banded electromagnetic waves and maintain amplitude and phase coherence across the target.

It is the purpose of the following chapter to analyze a set of practical propagation media with a view to obtaining analytical expressions providing information about

- pathloss,
- bandwidth (longitudinal resolution capability), and
- spatial coherence (transverse resolution capability).

We shall endeavor to offer a unified set of theoretical expressions based on simple first-principle physics. The aim is to present theoretical expressions which lend themselves to further analyses, so as to form the basis for adaptive manipulations, and thus maximize system performance. Furthermore, we shall base our theoretical approach on the concepts derived in Chapter 2 above, where we presented the basic theory for scattering/diffraction.

Ideally, we would like to organize ourselves, to measure first the properties of the propagation medium, then the characteris-

54 ADAPTIVE RADAR

tics of the target background, and finally, knowing in advance the signature of the target, we could optimize the total system performance. This is illustrated in the form of an artist's conception in Figure 5.1.

LINE-OF-SIGHT PROPAGATION

Figure 5.2 shows the distance dependence of a radio field for various propagation mechanisms. With line-of-sight, the power density decreases with distance as R^2 and in terms of transmitted power P_T, gain of transmitting antenna G_T, area of receiving aperture A, and distance between transmitter and receiver R the received power P_R is given by:

$$\frac{P_R}{P_T} = \frac{G}{4\pi R^2} A \tag{5.1}$$

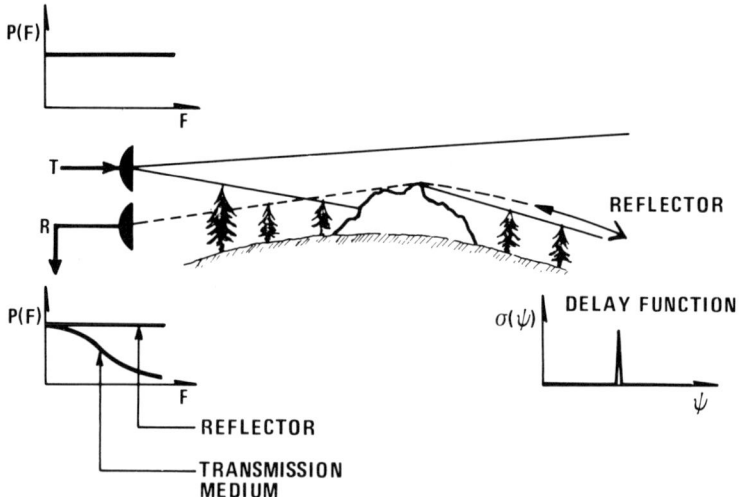

Figure 5.1 Limitations on radar performance imposed by the propagation medium. Having "test targets" at our disposal, we can assess the influence of the intervening propagation medium.

The received power decreases gradually as the distance increases; for a given power and beam configuration there is no parameter which can be altered to improve the situation.

Let us consider what happens if part of the transmitted energy illuminates the ground surface and is reflected into the direct wave. If H_T is the height above the reflecting surface of the transmitting antenna and H_R is the corresponding height of the receiving antenna, the difference in path length of the direct wave relative to that of the reflected wave is given by:

$$\Delta = \frac{2H_T H_R}{R}$$

Phase angle Φ between the two waves is therefore:

$$\Phi = \frac{4\pi H_T H_R}{\lambda R}$$

When $\Phi = \pi$ the two waves appear in antiphase, giving rise to a minimum in the plot of received power vs distance, as shown in Figure 5.2. Similarly, if the phase angle is equal to 2π, we have

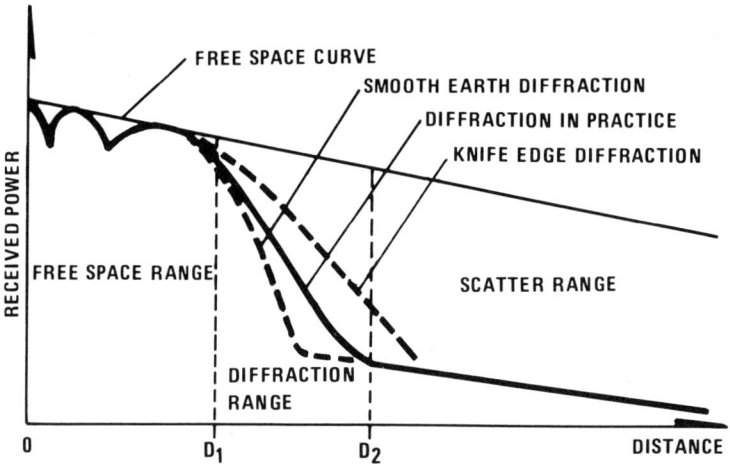

Figure 5.2 Factors determining the path loss in tropospheric propagation.

constructive interference and get a maximum. Accordingly, to change a maximum to a minimum without altering the geometry, i.e., the position of the transmitter and the receiver, the wavelength will have to be changed by a factor of 2. This, obviously, is within reach in an adaptive communication system (see below).

This statement is strictly correct only if we neglect the effects of surface waves. If this cannot be neglected we shall have to include the surface wave in addition to the direct wave and the reflected wave. This involves adding the surface wave term:

$$(1 - R) \, A \, e^{j\phi}$$

where R = the reflection coefficient of the ground
A = the surface wave attenuation factor

These parameters vary with polarization and the electrical constants of the ground. For near grazing paths, R is approximately equal to -1, and A can be neglected if the antenna is elevated more than a wavelength above the ground (or more than 5-10 wavelengths above sea water). This condition is generally fulfilled in practice when dealing with microwaves.

Now let us assume that we would like to adjust the height of the receiving antenna to achieve constructive interference. As seen from the above expression, this is achieved by shifting the receiving antenna vertically through a distance:

$$\Delta H_R = \frac{\lambda R}{4H_T} \qquad (5.2)$$

Hence, if we are dealing with a target of vertical extent ΔH_{TARGET} we shall have to control the propagation parameters to ensure that $\Delta H_R \gg \Delta H_{TARGET}$.

Then let us, in terms of bandwidth, consider the properties of a propagation circuit involving ground reflections. Having already calculated the delay function related to our transmission circuit, the bandwidth is readily obtained as a function which is proportional to the inverse of delay. Specifically, if the delay function, as in our case, consists of two δ functions with separation:

$$\Delta = \frac{2H_T H_R}{R}$$

The correlation properties in the frequency domain of the reflected signal is given by the Fourier transform of the delay function. The Fourier transform of two δ functions is a cosine relationship. The half-power bandwidth of the first order is, as we have already seen above, given by:

$$\Delta F = 0.25 \frac{c}{\Delta} \qquad (5.3)$$

Introducing the geometric expression for Δ above, we find that the bandwidth is given by

$$\Delta F = \frac{c \cdot R}{8 H_T H_R} \qquad (5.4)$$

The simple formulas above are based on the assumption that radio rays propagate along straight lines. If the atmosphere does not have a homogeneous refractive index, this is not the case. If we are dealing with a vertical profile of refractive index we shall experience bending (refraction). Knowing the refractive index profile, we can calculate the ray bending from Snell's law. We are thus able to calculate the total bending to which a ray is subjected when propagating from the radar to the target (and back along the same path). This bending is dependent on the initial direction ϕ of the ray relative to the isosurface of refractive index [32]. In this simple treatment we shall limit ourselves to nearly horizontal directions of the radar beam.

58 ADAPTIVE RADAR

From the simple geometry shown in Figure 5.3, we have the following geometric relationships:

$$\frac{1}{a} = \frac{1}{R} + \frac{dn}{dz}\cos\phi \tag{5.5}$$

$$a = kR$$

$$k = \frac{1}{1 + R\dfrac{dn}{dz}\cos\phi} \tag{5.6}$$

Direction to target for the case of no ray bending is given by:

$$\alpha_0 = \frac{d}{R}$$

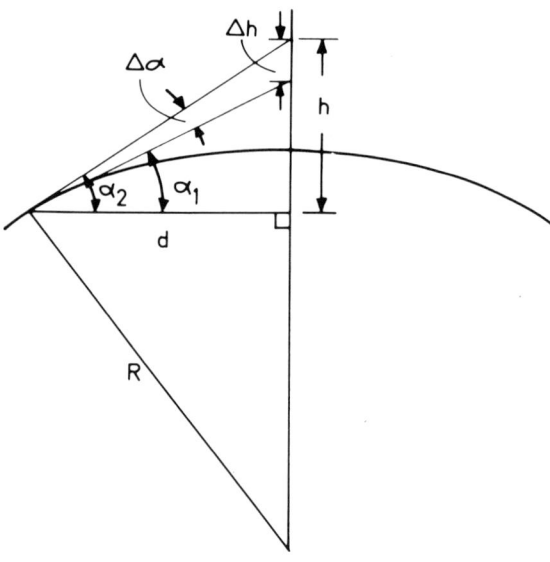

Figure 5.3 Geometry related to height errors caused by refraction.

R = the real earth radius. Similarly, with a bending corresponding to an effective earth radius a, the ray direction is:

$$\alpha_1 = \frac{d}{a}$$

i.e.,

$$\frac{\Delta h}{h} = \frac{(\alpha_0 - \alpha_1)d}{\alpha_1 d}$$

and

$$\frac{\Delta h}{h} = \frac{\alpha_0}{\alpha_1} - 1 = (k - 1) \tag{5.7}$$

where k is the ratio of effective earth's radius to real radius.

Figure 5.4 shows a practical example of the probability distribution of the bending parameter a/R. These were obtained from conventional routine meteorological radiosondes released from Sola, Norway, during the winter of 1965.

Based on the probability distribution of the effective to real earth radius, the total bending (height error) can be calculated. These are shown in Figure 5.5 for zero elevation angle. For the sake of illustrating the severity of height errors, numbers are given for the specific case of a target at a height of 1500 meters.

So far we have restricted the discussion to the case where the spatial small-scale fluctuations in refractivity were negligible; we shall thus experience refraction only and no scattering. We shall now consider an atmosphere which is characterized by small-scale spatial refractivity fluctuations and the effect of these on line-of-sight propagation.

As a consequence of the fact that irregularities in the atmospheric refractive index structure lead to multipath phenomena and delay variations when an electromagnetic wave passes through the irregular transmission medium, we suffer a loss in

60 ADAPTIVE RADAR

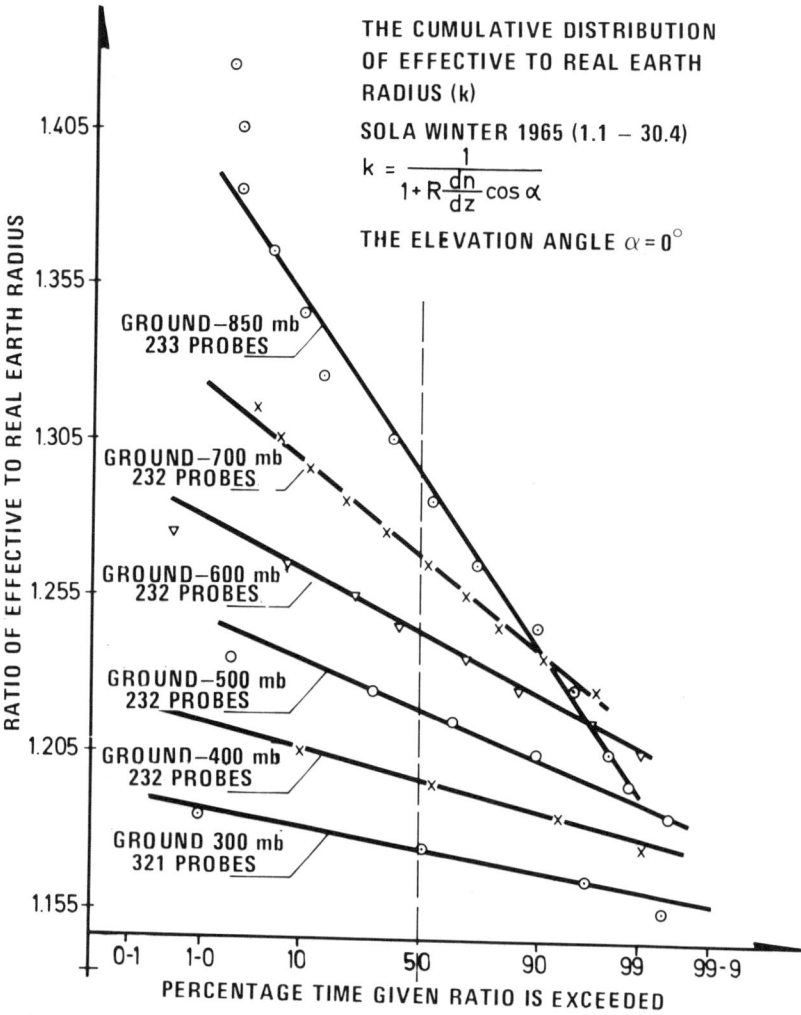

Figure 5.4 Probability distribution of the ratio of effective to real earth radius calculated from data from routine meteorological radiosondes.

bandwidth. We shall now give some theoretical results, which are well confirmed experimentally, giving information about the amplitude covariance as a function of frequency separation (i.e., bandwidth properties of the medium) and as a function of spatial separation.

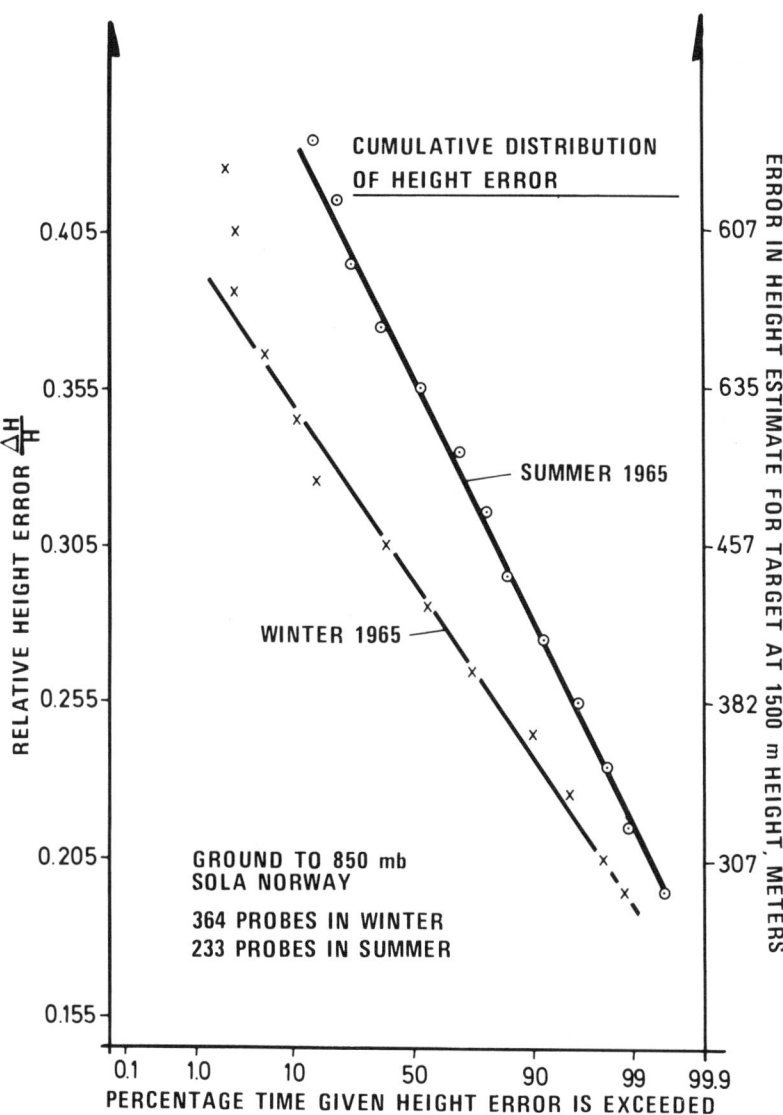

Figure 5.5 Probability distribution of height error calculated on the basis of radiosonde information presented in Figure 5.4.

Before referencing the results of comprehensive calculations, we shall, as above, give some quantitative results for the purpose of ensuring a physical understanding of the basic physics involved.

62 ADAPTIVE RADAR

Referring to the simple geometric sketch of Figure 5.6, we see that there are two extreme paths through which the electromagnetic waves can travel from the transmitter T to the receiver R within the first Fresnel zone. One is the shortest direct way from T to R, the other is via a path which is half a wavelength longer than the direct route. The result of these two waves is the vector sum of two signals with a 180° difference in phase, causing destructive interferences.

To the first order, therefore, we would expect the width θ of the angle of arrival spectrum at the receiving point to be given by

$$\tan^{-1} \theta = (\lambda d)^{1/2}/d$$

$$\theta \cong (\lambda/d)^{1/2}/d \tag{5.8}$$

Knowing the angular power spectrum (the beamwidth), the correlation distance in a plane through the location of the receiver normal to the line T–R can be calculated. We have shown above that this spatial correlation of field strength is the Fourier transform of this angular power spectrum.

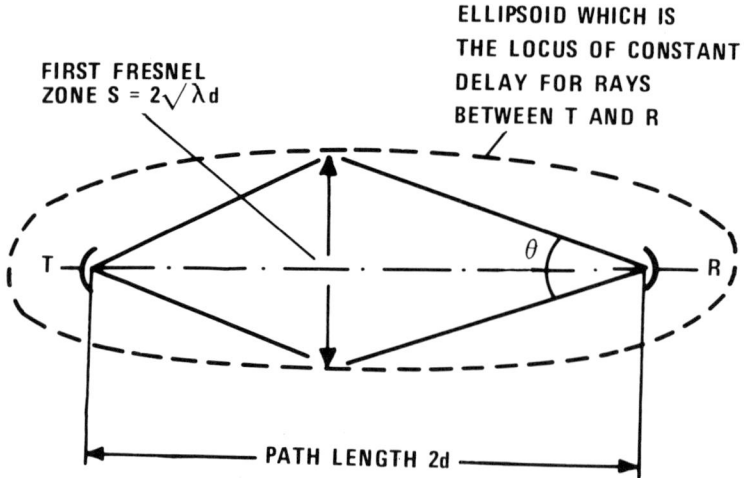

Figure 5.6 The geometry of line-of-sight propagation.

If this power spectrum is a (sin x)/s function, the Fourier transform is a rectangular function. If the width of this, i.e., the correlation distance of field strength, is L, we have the following relationship between the half-power width of the beam $\theta_{1/2}$ and the correlation distance L:

$$\theta_{1/2} = \frac{0.88 \lambda}{L} \tag{5.9}$$

where λ = the wavelength of the electromagnetic wave.

In passing, note that this is the same expression as that relating antenna beamwidth $\theta_{1/2}$ to the antenna aperture diameter L. This is not surprising since the antenna radiation pattern [the $P(\theta)$ function] is the Fourier transform of the illuminating field strength distribution over the antenna aperture. From Equations 5.8 and 5.9, therefore, we have

$$\frac{0.88 \lambda}{L} = (\lambda/d)^{1/2}/d$$

i.e., correlation distance

$$L = 0.88 (\lambda d)^{1/2} \tag{5.10}$$

Thus, the correlation distance of field strength transverse to the line of propagation is comparable with the first Fresnel zone. Then let us calculate the bandwidth $\Delta \omega$. We have learned that the bandwidth function (autocorrelation function in the frequency domain) is obtained by Fourier transforming the delay function.

For simplicity, let us again assume that the delay function is of the (sin x)/x form with a half-power width $\Delta \tau = (\lambda/2)/C$. The bandwidth function, i.e., the frequency transfer function P(F) or the correlation function in the frequency domain R(ΔF), would then be a rectangular function the width of which is

$$\Delta F = \frac{1}{\Delta \tau} = \frac{2C}{\lambda} \tag{5.11}$$

64 ADAPTIVE RADAR

These were the approximate results. Lee and Harp [33] have performed rigorous calculations based on a general expression for the spatial distribution of refractive index. For details the reader is referred to Lee and Harp [33] and Gjessing et al. [34].

Summing up this section on line-of-sight propagation mechanisms in relation to an adaptive radar (or communications) system, the following should be noted:

1. The pathloss can be minimized by adjusting the height of the transmitting antenna or the wavelength.
2. The vertical coherence distance of field strength can likewise be maximized by adjusting the antenna height: the lower the antenna height, the larger the coherence distance.
3. If we are dealing with a structure of homogeneous refractive index, the bandwidth of the line-of-sight circuit can be optimized by decreasing the height of the transmitting antenna.
4. Spatial fluctuations in refractive index resulting from atmospheric turbulence impose severe limitations on bandwidth as well as on the vertical coherence distance. There is nothing we can do to improve this situation, except noting that the turbulent transmission medium is very variable. Given sufficient time, there will be a time interval where conditions are very much better than the average conditions.

PROPAGATION MECHANISMS INVOLVING SCATTERING AND DIFFRACTION

As the demand for reliable broad-band communication circuits, high-resolution adaptive radar systems increases, so does the need for detailed information about the transmission medium.

The multitude of new demands leads to a versatile and sophisticated usage of the transmission medium. This, in turn, calls for a very comprehensive description of the medium. The transmission medium constitutes the limiting factor in many interesting and potentially powerful communication and radar techniques of which the following should be mentioned:

1. Spread spectrum modulation systems and multifrequency radar systems are very sensitive to frequency selective fading, i.e., require a large instantaneous bandwidth. Hence we shall need information about the circuit bandwidth and its variability.
2. Large synchronous time division multiplex communication systems set narrow limits with regard to variations in time delay in the system. As a consequence, information about the delay spectrum and its temporal variability is mandatory.
3. Environmental surveillance systems requiring a large spectrum of wavelengths to illuminate the scene of interest. To optimize system performance, we shall need detailed information about the effects of the intervening propagation medium.

With these applications particularly in mind, the next sections will discuss scattering mechanisms and channel characterization in relation to broad-band communications and multifrequency, high-resolution radar systems.

Basic Relationships in Over-the-Horizon Scatter Propagation

When discussing the characteristic properties of a scattered (or diffracted) wave in relation to radar and communication systems, it is useful to have a physical understanding of the basic principles involved. With reference to previous discussion and earlier works [1,10,35], a brief sketch of some of the more important derivations will be given.

Consider a volume element $dv = dx\,dy\,dx = d^3\vec{r}$ within the scattering volume V, this scattering volume being confined to the spatial region in the troposphere illuminated by the transmitting antenna and "seen" by the receiving antenna. If the permittivity (refractive index squared) within the elementary volume differs by an amount $\Delta\epsilon$ from the average value of the permittivity ϵ_0, the element of dielectric becomes polarized, giving rise to a dipole moment $d\vec{P} = \Delta\epsilon\,dv\,\vec{E}_0$ when under the influence of an electric field $\vec{E}_0$. At distance R from the scattering element, the dipole moment results in a polarization potential $d\Pi$ and, provided $k^2\vec{\Pi} \gg \nabla\nabla\cdot\vec{\Pi}$ (which requires $R \gg V^{1/3}$), the scattered field strength $\vec{E}_s = k^2\vec{\Pi}$, where $\vec{k}$ is the wave number of the electric field. The scattered field resulting from the integral of elementary scattering elements is then given by:

$$E_s = \frac{k_s^2}{4\pi R} \int E_0(\vec{r})\,\epsilon(\vec{r},t)\,e^{-j\vec{K}\cdot\vec{r}}\,d^3\vec{r} \qquad (5.12)$$

where $\vec{K} = \vec{k}_0 - \vec{k}_s$
$\vec{k}_0, \vec{k}_s$ = the wave numbers of the incident and the scattered fields, respectively, such that $|K| = (4\pi/\lambda)\sin\theta/2$, where θ, the scattering angle, is the angle between $\vec{k}_s$ and $\vec{k}_0$.

Note that Equation 5.12, which is derived from Maxwell's equation, is perfectly general and does not consider the nature of the refractive-index irregularities described by the function $\epsilon(\vec{r},t)$. This function may be a stochastic one, in which case the refractive-index field is conveniently described by the spatial autocorrelation function of refractive index, or we may be dealing with an ordered variation in ϵ, say a horizontal layer through which the refractive index varies in a systematic fashion expressible as a well-behaved function (see below).

In exactly the same way, $\vec{E}_0(\vec{r})$ describes the spatial variations in the electric field within the scattering volume. From the basic equation above, we see that there are two limiting cases: First, if the field $\vec{E}_0(\vec{r})$ is constant within the scattering volume (or varies slowly in space in comparison with the $\epsilon(\vec{r})$ function),

$$\vec{E}_s = \frac{k^2 \vec{E}}{4\pi R} \int_V \epsilon(\vec{r}) e^{-j\vec{K} \cdot \vec{r}} d^3 \vec{r}$$

which states that the scattered field $\vec{E}_s$ is proportional to the Fourier transform of the spatial variation in refractive index within the scattering volume V. Second, if $\epsilon(\vec{r})$ is constant within the scattering volume, the equation tells us that the diffraction field $\vec{E}_D$ is the Fourier transform of the spatial variation in field strength within the scattering volume V (see below):

$$\vec{E}_D = \frac{k^2 \epsilon}{4\pi R} \int_V \vec{E}(\vec{r}) e^{-j\vec{K} \cdot \vec{r}} d^3 \vec{r}$$

The intermediate conclusions are visualized in Figure 5.7.

Now let us concentrate on the scattered field associated with spatial variations in refractive index $\epsilon(r)$. We shall need information about the scattered power P as a function of scattering

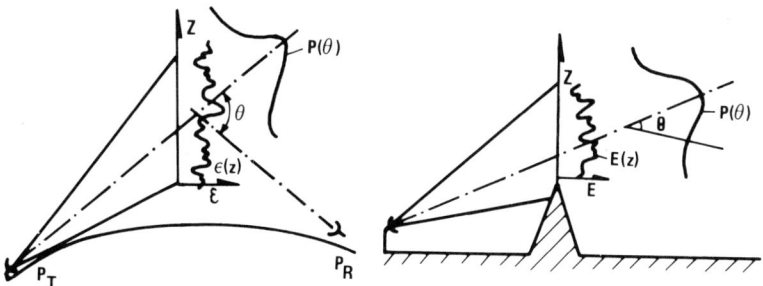

Figure 5.7 (left) The scatter field is the Fourier transform of the spatial variations in refractive index $\epsilon(z)$; (right) the diffracted field is the Fourier transform of the spatial variation in the illuminating field $E(z)$.

angle θ. To obtain this, we multiply $\vec{E}_s$ by its complex conjugate $\vec{E}_s^*$ obtaining the scattering cross section σ given by:

$$\sigma(\theta) = (\pi k^4/2)\, \Phi(\vec{K}) \qquad (5.13)$$

where $\Phi(\vec{K})$ = the spatial "power spectrum" of the refractive-index irregularities such that $\Phi(\vec{K})$ is the Fourier transform of the spatial autocorrelation function of $\epsilon(\vec{r})$.

The scattering cross section σ is defined as the mean power in the scattered wave per unit power density of the incident wave in the scattering volume, per unit solid angle in the direction of $\vec{k}_s$, per unit scattering volume. Note that the power spectrum $\Phi(\vec{K})$ is the Fourier transform of the spatial autocorrelation function of refractive index fluctuation.

Having obtained an expression for the angular power spectrum of the scattered field in terms of the function $\Phi(\vec{K})$ describing the spatial variation in refractive index, the questions which arise are:

- Does there exist a unique form of the $\Phi(\vec{K})$ function for the tropospheric propagation medium?
- To what extent does the $\Phi(\vec{K})$ function vary with time and geographic location of the scattering volume?

Many forms of the $\Phi(\vec{K})$ function have been suggested. The more important ones are associated with the following names: Oboukov-Kolmogorov, Booker-Gordon, Bolgiano, Willers-Veisskopf and Norton.

In this brief discussion of the subject, a detailed discussion of the relative merits and justification for the various $\Phi(\vec{K})$ functions does not seem justified. There is good justification [10,36] for writing the power spectrum in the form

$$\Phi(\vec{K}) \cong K^{-n} \qquad (5.14)$$

where n = a number that, depending on the atmospheric conditions, may vary between approximately 2 and 7

Many theories predict n = 11/3. Experiments show [37–43] that n varies within wide limits. The theoretical value n = 11/3 appears to be close to the median value of the observations (see below).

Based on the refractive index spectrum expressed in the form $\Phi(K) \cong K^{-n}$, we shall now calculate some of the characteristic parameters of a long-distance forward scatter circuit. There are many that should be mentioned: time-delay spectrum, bandwidth, horizontal and vertical correlation distance of field strength, and antenna-to-medium coupling loss. These will now be considered.

Calculation of Pulse Distortion in Terms of Radiometeorological Parameters

Our task is now to calculate the delay spectrum (pulse distortion) and subsequently the bandwidth on the basis of information about the refractive index structure $\epsilon(\vec{r})$ as expressed by its spatial power spectrum $\Phi(K)$, written in the form $\Phi(K) \cong K^{-n}$.

Using a wide-beam antenna so that the multipath transmission is governed by the scattering mechanism rather than by the beam geometry, we first seek an expression relating path length ℓ and the position in space of the scattering element, i.e., we require an expression relating ℓ and the scattering angle θ (Figure 5.8). If d is the length of the chord between T and R, simple geometric calculations give the required results:

$$\theta = 2 |(\ell/d)^2 - 1|^{1/2}$$

If we transmit a short radio pulse, the power that reaches the receiver has traveled through a wide variety of different paths.

By substituting for θ in the expression for the angular power spectrum ($P \sim \theta^{-n}$), we get the spectrum relating power and path length. Normalizing this power with respect to the power

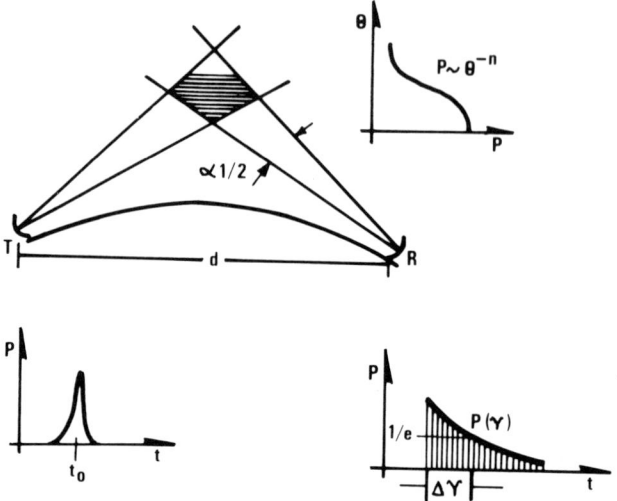

Figure 5.8 The delay spectrum is determined by the variation in path length.

received via the shortest propagation path ℓ, namely, that determined by the earth's tangent planes, we get:

$$\ell_0 = |d|\ell + (d/2a)^2|^{1/2}$$

where a = the effective earth radius

The power spectrum takes the form:

$$|P(\ell)/P(\ell_0)| = \left(\frac{4a^2}{d^2}\right)^{-n/2} |(\ell/d) - 1|^{-n/2}$$

Expressing this spectrum in terms of the path length $\Delta\ell$, which is in excess of the minimum path length ℓ_0 (i.e., writing $\ell = \ell_0 + \Delta\ell$), we find that the power spectrum referred to ℓ_0 is given by:

$$|P(\Delta\ell)/P(\ell_0)| = |1 + (8a^2/d^3)\Delta\ell|^{-n/2}$$

RADIO PROPAGATION THROUGH THE ATMOSPHERE

Since $\Delta \ell = \tau C$, where C is the velocity of light, the delay spectrum referred to τ_0 (the shortest time delay) is given by:

$$| P(\tau)/P(0) | = | 1 + (8a^2 \tau C/d^3)|^{-n/2} \qquad (5.15)$$

The 1/e width of this delay spectrum is then given by:

$$\Delta \tau = (d^3/8a^2 C)(e^{2/n} - 1) \qquad (5.16)$$

If, on the other hand, the antenna beams are narrow, such that the spread in the path length is determined by beam geometry rather than by the $\Phi(K)$ function (i.e., the beams are so narrow that $\Phi(K)$ can be considered constant when θ varies within the scattering volume), we can find a simple expression for the delay spectrum in terms of geometric parameters. Under these conditions the delay spectrum will have a width $\Delta \ell/C$ given by [44]:

$$\Delta \tau = d/2C \cdot [(d/a)\beta + \beta^2]$$

where β = the beamwidth

Having expressed the delay spectrum in terms of radiometeorological parameters, we shall now apply a similar method to calculate the bandwidth of the transmission channel.

Calculation of Bandwidth

To avoid the confusion which often arises when the term "bandwidth" is used in relation to scattering processes, let us define bandwidth.

Consider the case where several radio waves having different frequencies are transmitted simultaneously. At the receiver, the power at each of these frequencies is measured as a function of time (i.e., we measure $P_{F_1}(t)$, $P_{F_2}(t)$, etc.). If we take the

72 ADAPTIVE RADAR

instantaneous ratio of power at the various frequencies and integrate the ratio:

$$\int \frac{P_{F_1}(t)}{P_{F_2}(t)} dt$$

we get information about the bandwidth.

On the other hand, if we integrate the signal at either frequency over the appropriate time interval by forming:

$$\frac{\int P_{F_1}(t) \, dt}{\int P_{F_2}(t) \, dt}$$

We do not obtain information about bandwidth, but something often referred to as "the wavelength dependence of the scatter circuit."

Forming

$$\int \frac{P_{F_1}(t)}{P_{F_2}(t)} dt$$

is one way of obtaining information about bandwidth. Another is the following.

As in the case above we transmit a set of radio waves having different frequency. We now make sure that all of these frequencies are correlated in amplitude and phase. This is, as an example, achieved by amplitude-modulating a carrier, thus obtaining two sidebands $2f_{AM}$ apart, if f_{AM} is the frequency of the modulating wave. These sidebands, obviously, are correlated in amplitude and phase. At the receiving end we pick up the two sidebands and correlate one with the other (i.e., we form the cross-correlation function $R_{12}(\tau)$. The more narrow-banded the transmission channel, the poorer is the correlation. (By transmitting many correlated waves spread over a frequency band, we can find the complete autocorrelation function $R(\Delta F)$

in the frequency domain as discussed above.) This is analogous, as we shall see in the following, to the power spectrum of the transmission channel.

By analyzing the first alternative first, we can use the results of simple network theory to obtain a simple approximate result. We know that the response to a delta pulse of a network is known as the impulse response $V(\tau)$ of the network. Furthermore, the Fourier transform of the impulse response is known as the transfer function $F(\omega)$ of the network. By multiplying this transfer function with its complex conjugate, we obtain the power spectrum that we are seeking. From the previous section we obtain the expression for the impulse response by taking the square root of Equation 5.15. Analyzing this function, we find that it closely resembles an exponential function of the form $P(\tau) = \exp(-\alpha\tau)$. The $1/e$ width of the impulse response is given by:

$$\Delta\tau = d^3 (e^{4/n} - 1)/8a^2 C \tag{5.17}$$

To simplify the Fourier transformation, we assume an exponential impulse response such that $\Delta\tau = 1/\alpha$ (when we introduce only a small error). The Fourier transform of the exponential impulse response $\exp(-\alpha\tau)$ is given by:

$$F(\omega) = (\alpha + j\omega)^{-1}$$

The power spectrum is then given by:

$$W(\omega) = F(\omega) F^*(\omega)$$
$$= (\alpha^2 + \omega^2)^{-1}$$

Substituting for α as obtained from Equation 5.17 and normalizing the resulting equation for $\omega = 0$, we find that the half-power width of the power spectrum is given by:

$$\Delta\omega = 8a^2 C d^{-3} (e^{4/n} - 1)^{-1} \tag{5.18}$$

74 ADAPTIVE RADAR

Now let us compute the autocorrelation function in the frequency domain $R(\Delta\omega)$. The voltage V_1 at frequency ω is given by $V_1 = F(\omega) = (\alpha + j\omega)^{-1}$. Similarly, the voltage V_2 at frequency $(\omega + \Delta\omega)$ is given by $V_2 = F(\omega + \Delta\omega) = [\alpha + j(\omega + \Delta\omega)]^{-1}$. The normalized complex autocorrelation of these two voltages is then given by:

$$R(\Delta\omega) = \frac{\int_{-\infty}^{\infty} [1/(\alpha + j\omega)] \{1/[\alpha - j(\omega + \Delta\omega)]\} \, d\omega}{\int_{-\infty}^{\infty} [1/(\alpha^2 + \omega^2)] \, d\omega}$$

By solving this integral we get the following expression for the modulus of the autocorrelation function:

$$R(\Delta\omega) = |1 + (\Delta\omega/2\alpha)^2|^{-\frac{1}{2}} \qquad (5.19)$$

The width of this autocorrelation function is obtained by letting $R(\Delta\omega) = \frac{1}{2}$, thus obtaining:

$$\Delta\omega_{\frac{1}{2}} = 16 \cdot 3^{\frac{1}{2}} a^2 \, Cd^{-3} \, (e^{4/n} - 1)^{-1} \qquad (5.20)$$

Note that the width of the autocorrelation in the frequency domain is $2(3)^{\frac{1}{2}}$ times the half-power width of the power spectrum.

Correlation Distance of Field Strength

In this section attention is focused on the spatial field-strength correlation properties of a scattered radio wave. We have a wide-beam transmitter radiating its power essentially in a horizontal direction. The scattered wave resulting from this transmitter impinges on two nearly identical, small-aperture receiving antennas positioned beyond the horizon relative to the transmitter. The antennas are spaced vertically or horizontally such that the center line through the receiving antennas is normal to the center line through the transmitter T and the receivers R. We measure the normalized complex correlation of the voltages induced in the antennas.

As shown above, this spatial field-strength correlation function is the Fourier transform of the angular power spectrum of the wave reaching the receiving antennas. Specifically, if the antennas are spaced horizontally, thus giving us information about the correlation properties of field strength along a horizontal direction, this correlation function is determined by the angle-of-arrival spectrum as measured in a horizontal plane. If the antennas are spaced vertically, it is the angle-of-arrival spectrum in the vertical plane that matters. Specifying the angle of arrival of a particular scattered wave by an elevation angle α (relative to the center line through T and R) and an azimuth angle β (relative to a great circle plane through T and R) we have in the same manner as above that a refractive-index spectrum $\Phi(K) \sim K^{-n}$ gives rise to an angular power spectrum of the form:

$$P(\alpha,\beta) \sim (\beta^2 + \alpha^2)^{-n/2} \tag{5.21}$$

The horizontal correlation of field strength is thus obtained by a Fourier transformation of P with respect to β, whereas the vertical correlation is obtained from the $P(\alpha)$ relationship.

A rigorous Fourier transformation of Equation 5.21, however, lends itself to numerical computations only. In our case, we need a simple approximate expression. This can be obtained if we approximate $P(\alpha)$ and $P(\beta)$ by a $(\sin \chi)/\chi$ function, thus giving us a simple expression for the Fourier transform. This is a procedure well known in antenna theory. From antenna theory we know that if L is the width of the illuminating field-strength distribution, then half-power width of the resulting angular power spectrum (beam width) is given by:

$$\theta_{1/2} = 0.88 \lambda/L \tag{5.22}$$

where λ = the radiowavelength

76 ADAPTIVE RADAR

By applying these results to our problem, we find that the 3-dB width of the scattered beam as measured in the vertical plane is given by:

$$P_{1/2}/P_o = 1/2 = (\alpha_o + \alpha_{1/2}/\alpha_o)^{-n} \qquad (5.23)$$

where $\alpha_{1/2}$ = the 3-dB beamwidth of the scattered beam
α_o = d/2a
d = the path length
a = the earth's radius.

By solving for $\alpha_{1/2}$ and substituting this in Equation 5.22, we find that the vertical correlation distance of field strength is given by:

$$L_v/\lambda = 0.44\,(a/d)\,(2^{1/n} - 1)^{-1} \qquad (5.24)$$

Similarly, the horizontal correlation distance is given by:

$$L_H/\lambda = 0.44\,(a/d)\,(4^{1/n} - 1)^{-1/2} \qquad (5.25)$$

These approximate expressions are in very good agreement with the results based on rigorous numerical transformations of the scattered angular power spectra [45].

Note that the correlation distance is only very weakly related to the spectrum slope, and that refraction effects play a dominating role.

Antenna Gain Degradation

From basic antenna theory we know that the free-space antenna gain is proportional to the antenna aperture: $G = 4\pi A \lambda^{-2}$. When dealing with large antennas in connection with scatter propagation, however, this linear relationship no longer holds.

If the antenna aperture is increased by a factor k, the received power is generally increased by a factor that is less than k. This apparent gain degradation is commonly referred to as antenna-to-medium coupling loss. The phenomenon can be explained in several different ways. We may base the discussion either on the width of the angular power spectrum of the scattered wave relative to the angular-reception capability of the receiving antenna, or on the spatial correlation distance of the received scattered field strength relative to the dimension of the antenna aperture. In this presentation we shall use the latter method for the purpose of illustrating the principles involved and to get an expression relating the gain degradation to the slope n of the refractive-index spectrum, rather than to seek an expression of optimum accuracy.

By basing our computations on the results of the previous section, we note that at the receiving site, the area (normal to the direction of propagation) over which the field strength is correlated is given by $L_v L_H$. This area may then be considered as being the effective receiving antenna aperture, provided that the actual aperture is larger than $L_v L_H$. If the actual area $A < L_v L_H$, we do not experience a gain degradation.

The effective antenna gain is thus:

$$G_{eff} = 4\pi L_v L_H \lambda^{-2} \qquad (5.26)$$

whereas the plane-wave gain is:

$$G = 4\pi A \lambda^{-2}$$

The gain loss is thus:

$$G_L = \frac{A}{L_v L_H} = \frac{5(2^{1/n} - 1)(4^{1/n} - 1)^{1/2} A}{(a/d)^2 \lambda^2} \qquad (5.27)$$

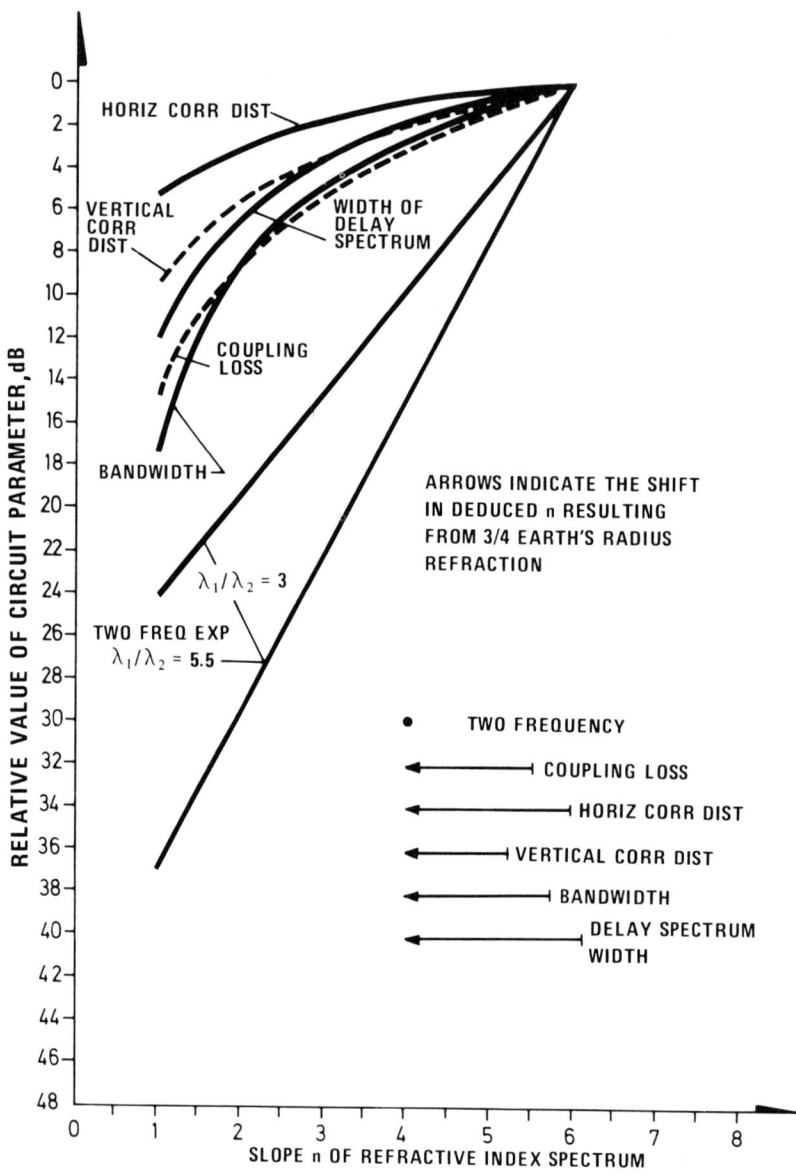

Figure 5.9 Theoretical relationship between the circuit parameter and the slope n of the refractive-index spectrum ($\Phi(K) \cong K^{-n}$). The curves show the degree to which the spectrum slope affects the circuit parameter.

Figure 5.9 shows the gain loss plotted against the refractive-index spectrum slope n. It should be emphasized that Equation 5.27 is based on the assumption of isotropy, as are indeed all the expressions for the circuit parameter.

If the atmospheric structure is horizontally layered (strongly anisotropic), the spectrum slope n associated with K vertical will be very much different from that associated with K horizontal. Therefore, to calculate the gain loss from Equation 5.27, we shall have to use different values of n for the two directions.

Wavelength Dependence of Scattered Power

Consider now the experiment involving simultaneous transmission and reception on two widely separated frequencies and scaled antennas. If the antenna beams are narrow such that the scattering volume is determined by the beam geometry rather than the scattering mechanism, the scattering volumes for the two frequencies are identical. Accordingly, the ratio of the power received on the two frequencies is given by the ratio of the corresponding scattering cross sections as presented in Equation 5.13. In this case, therefore, the wavelength dependence is the ratio of power received on the two frequencies given by:

$$P(\lambda_1)/P(\lambda_2) = (\lambda_1/\lambda_2)^{n-2} \qquad (5.28)$$

In Figure 5.9, the power ratio is plotted logarithmically to the basis of n and normalized for n = 6.

Two different wavelength ratios are used, namely, $\lambda_1/\lambda_2 = 3$ and $\lambda_1/\lambda_2 = 5.5$. If for some practical reasons (e.g., ground reflections) the effective gain of the two scaled antennas are not exactly identical, error is introduced. It can be shown, however, that the error is proportional to the square root of the antenna gain ratio only. Note that the term wavelength dependence often refers to the case where the received power

80 ADAPTIVE RADAR

is normalized with respect to free-space transmission loss. This normalized power ratio then takes the form:

$$[(P/P_{FS})(\lambda_1)]/[(P/P_{FS})(\lambda_2)] = (\lambda_1/\lambda_2)^{n-4} \qquad (5.29)$$

Summing up the section on circuit parameters in relation to scatter propagation in terms of radiometeorological parameters n (refractive index irregularity spectrum) and a (effective earth radius), Table 5.1 and Figure 5.9 are presented.

Table 5.1 Some Relationships Characterizing a Communication Channel

Communication Circuit Parameter	Relation Between Circuit Parameter and Radiometeorological Parameter
Width of delay spectrum	$\Delta\tau = \dfrac{d^3}{8a^2 c}(e^{2/n} - 1)$
Bandwidth	$\Delta\omega = \dfrac{8a^2 c}{d^3}(e^{4/n} - 1)^{-1}$
Gain loss	$G_L = \dfrac{5Ad^2}{\lambda_a^2}(2^{1/n} - 1)(4^{1/n} - 1)^{1/2}$
Horizontal field-strength correlation distance	$L_H = \dfrac{0.44 \lambda a}{d(4^{1/n} - 1)^{1/2}}$
Vertical field-strength correlation distance	$L_V = \dfrac{0.44 \lambda_a}{d(2^{1/n} - 1)}$
Wavelength	$\dfrac{P(\lambda_1)}{P(\lambda_2)} = (\lambda_1/\lambda_2)^{n-2}$

Radiometeorological Parameters n and a in Relation to Routine Meteorological Observations

This topic has been considered in some degree of detail in two earlier publications [34,35]. We shall give a summary of the results here. As we have already mentioned, and as will be substantiated in the following section, there are two radiometeorological parameters which are of dominating importance with respect to the characteristic properties of a forward-scatter circuit. One is the effective earth radius a; the other is the spectrum slope n of the refractive index irregularity spectrum. We shall now discuss the relationship between purely meteorological factors and the parameters a and n.

Determination of Effective Earth Radius a from Radiosonde Measurements

The refractivity N (where $N = (n - 1) 10^6$, n being the refractive index) is obtained from meteorological parameters by the Debye relationship:

$$N = 77.6 \, P/T + 3.73 \times 10^5 \, (e/T^2) \qquad (5.30)$$

where P = the total pressure (mbar)
          T = the absolute temperature
          e = the water vapor pressure in (mbar).

Thus from knowledge about the vertical profile of P, T and e as obtained from a conventional radiosonde, we can calculate the refractivity profile.

Knowing the N profile, we can calculate the ray bending from Snell's law (see above). We are thus able to calculate the total bending to which a ray is subjected, when propagating from the transmitter to the center of the scattering volume.

Similarly, we can calculate the bending experienced from the midpath point to the receiver. (In practice, we perform the calculation from the receiver back to the midpath point). We then know the effective scattering angle, which is the difference between the angle between the earth tangent planes through the transmitter and receiver and the total ray bending.

If d is the distance between the transmitter and receiver, the angle θ between the tangent planes is given by:

$$\theta = \frac{d}{R}$$

where R = the real earth radius

By the same relationship we then obtain the effective earth radius a using the effective scattering angle.

Alternatively, if the refractivity gradient dN/dz is constant through the height interval involved (from ground to scattering volume), the effective earth radius is given by the simple relationship:

$$\frac{1}{a} = \frac{1}{R} + \frac{dN}{dz} \times 10^{-6}$$

assuming a nearly horizontal direction of the radio beam.

We observe, then, that on the basis of P, T and e data from a conventional radiosonde ascent, we can obtain the radiometeorological parameter a appropriate for a given height of the scattering volume (corresponding to a given path length). The solid lines in Figure 5.10 show probability distribution of the ratio a/R based on 230 radio soundings at Sola in southwestern Norway during 1966 [35].

For comparison, the dashed line shows the ratio a/R based on 45 radiosonde ascents during October and November 1970 at Maniwaki, near Ottawa, Canada. This line lies intermediate to the Sola curves, but has a slope similar to the Norwegian summer data.

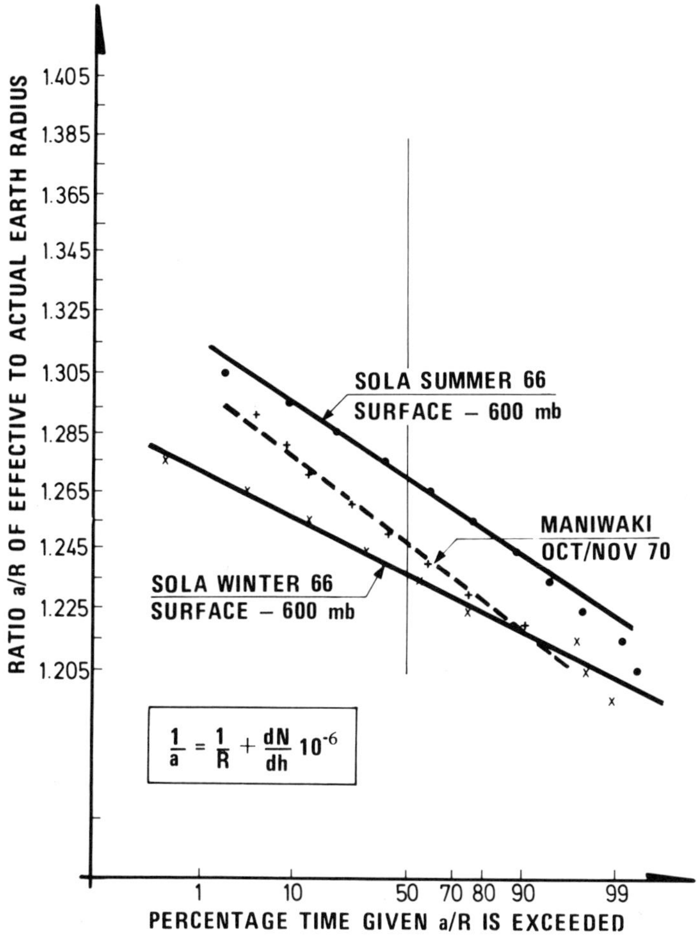

Figure 5.10 Distributions of the ratio of effective to actual earth radius, based on radiosonde observations at Sola, Norway, and Maniwaki, Canada.

*Determination of Spectrum Slope n
from Radiosonde Measurements*

Here the reader is referred to Gjessing et al. [34] where an empirical relationship was found between the atmospheric stability (a somewhat modified version of the well-known Väisälä-Brunt frequency v^2) and the slope n of the refractive index irregularity spectrum. The conventional Väisälä-Brunt frequency appears as the numerator in the Richardson's number and is normally written as:

$$v^2 = \frac{g}{T}\left(\frac{dT}{dz} + \frac{g}{c_p}\right) \qquad (5.31)$$

where g = the gravitational constant
　　　　T = temperature
　　dT/dz = vertical temperature lapse rate
　　　　c_p = specific heat at constant pressure

We see that the expression within the parentheses is a measure of the difference between the actual temperature lapse rate and the adiabatic lapse rate. This expression normally refers to a limited vertical section of the atmosphere.

The correlation of the spectrum slope n with atmospheric parameters describing the dynamic state of the atmosphere was significantly improved when a particular weighting function was placed on the temperature contribution to the stability.

Specifically, by adding a number which is determined by the temperature at the 850-mbar surface (1500 m altitude) to the Väisälä-Brunt frequency, the n - v^2 correlation was improved. For our particular purpose therefore, a modified version of v^2 was used:

$$v^2 = \frac{g}{T}\left(\frac{dT}{dz} + 5.50 \times 10^{-3}\, T_{850} + \frac{g}{c_p}\right) \qquad (5.32)$$

Here T (°K) and dT/dz (°K/100m) are average values obtained over the 850- to 400-mbar levels (1.5– 7 km altitude).

Having obtained v^2 from the results of a temperature profile determined by a conventional radiosonde observation, the spectrum slope n is found, using the expression for the n versus v^2 regression line:

$$n = 40.6 + 708\, v^2 \qquad (5.33)$$

Figure 5.11 shows probability distributions of spectrum slope n. The two upper curves are for the radiosonde station Sola in southwestern Norway for summer and winter 1966. These curves correspond to the distributions of Figure 5.10, and form the basis for the calculations to be presented in the following sections. For comparison, Figure 5.11 also shows n distributions for Maniwaki and Aalborg, northern Denmark, as obtained on the basis of radiosondes. The "predicted" n distribution for Aalborg is compared with that deduced from radio beam-swing experiments [10,37].

*Channel Characterization Statistics
on the Basis of Meteorological Data*

From the preceding chapter, we are in a position to calculate the probability distributions for some of the important transmission channel parameters. From the meteorological data as converted to the radiometeorological parameters n and a and presented in Figures 5.10 and 5.11, we are able to calculate channel parameters such as pulse distortion and bandwidth from the list of basic expressions given in Table 5.1.

A set of such probability distributions is given in Figures 5.12 to 5.17 (for further details on other channel characterization parameters, the reader is referred to Gjessing [46]).

Scattering from Atmospheric Layers (Waves)

In the previous chapter, the theory was based on the assumption that the refractive index irregularities were distributed

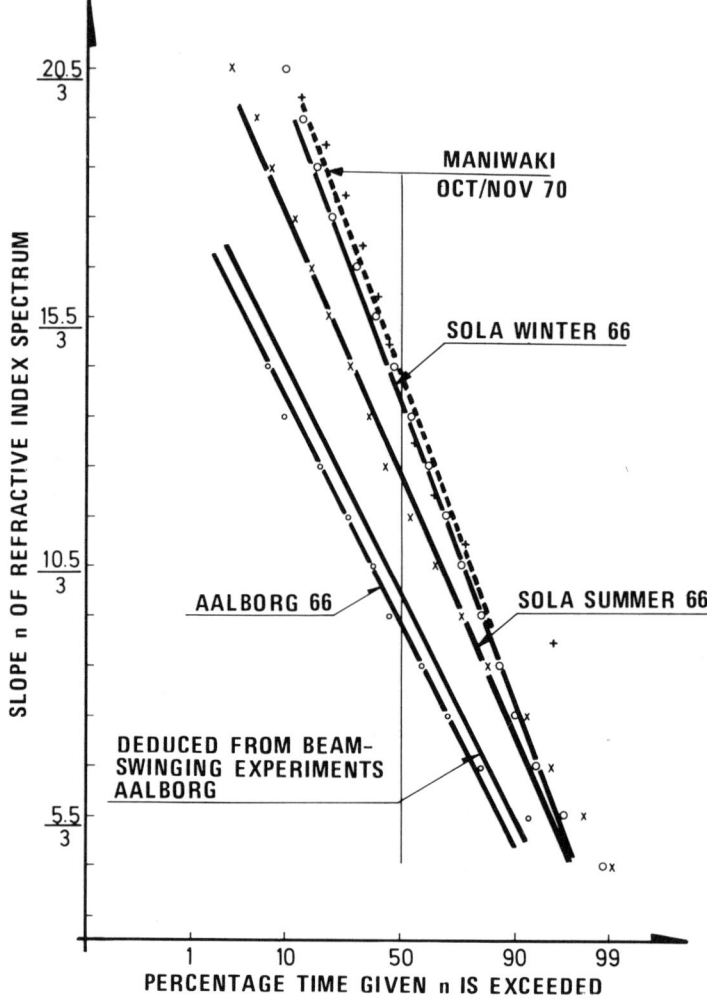

Figure 5.11 Distributions of the slope of the spectrum of refractive index irregularities as deduced from radiosonde observations and beam-swinging experiments.

randomly in the atmosphere, and superimposed on a gradual decrease of refractive index with height. We shall now consider another extreme case where the scattering/reflection is largely a result of a thin strata through which the refractive index varies drastically. For the purpose of illustrating the physics and the

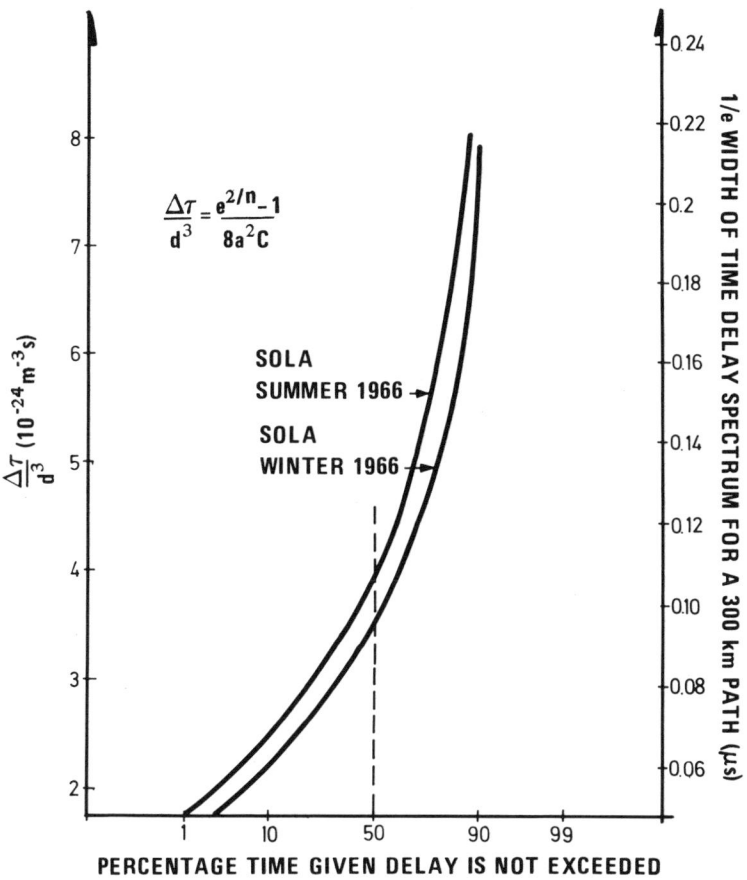

Figure 5.12 Width $\Delta\tau$ of the time delay function.

fundamental principles of the problem at hand, we shall explicitly present three different idealized atmospheric structures. These are shown in Figure 5.18. Here the first sketch refers to the case we have discussed in the previous section. The middle sketch will now be considered. Here we assume an irregular refractive index profile characterized by a layer of thickness Δh, through which refractive index varies linearly by total amount Δn.

88 ADAPTIVE RADAR

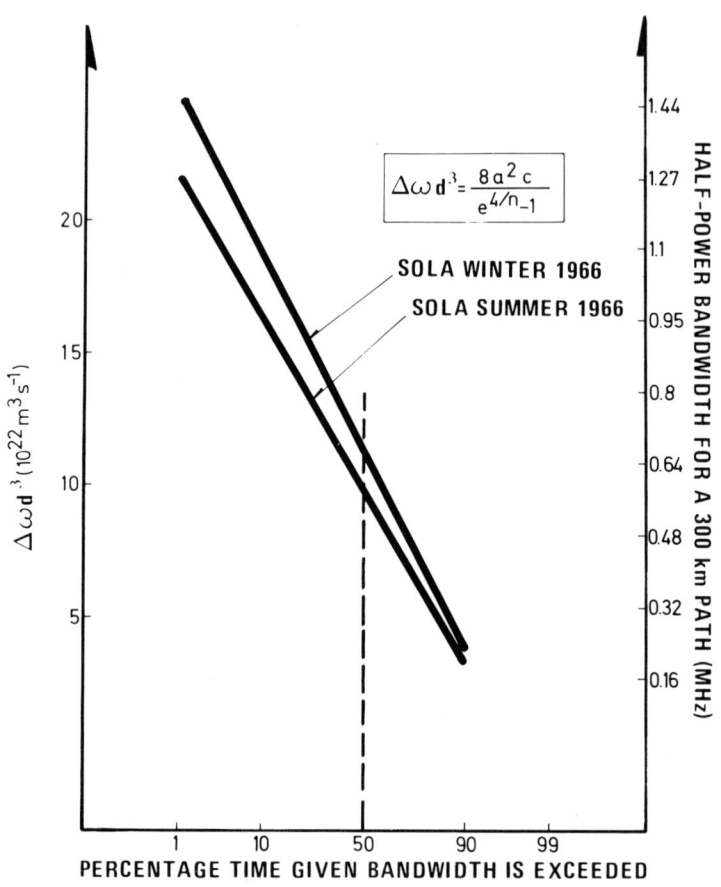

Figure 5.13 Half-power bandwidth $\Delta\omega$.

As we have already inferred above, and treated in detail in Chapter 2, the angular power spectrum of the scattered (reflected) wave is obtained by a simple Fourier transformation of the refractive index profile within the scattering volume. This procedure has been substantiated by several authors [47]:

$$|\sigma|^2 = \left| \frac{\Delta n}{2 \sin^2 \frac{\theta}{2}} \left(\frac{\sin X}{X} \right) \right|^2 \qquad (5.34)$$

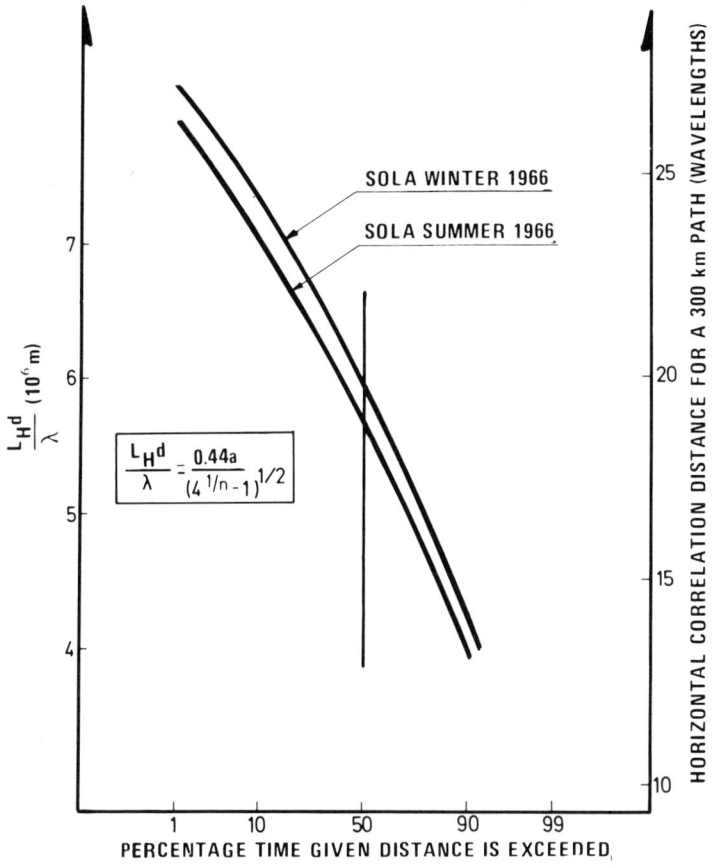

Figure 5.14 Horizontal field strength correlation distance.

Here X is a function of the geometry as follows

$$X = \frac{2\pi \Delta h}{\lambda} \sin \theta/2$$

where θ is the scattering angle.

This function is plotted to the basis of scattering angle for various layer thicknesses in Figure 5.19.

90 ADAPTIVE RADAR

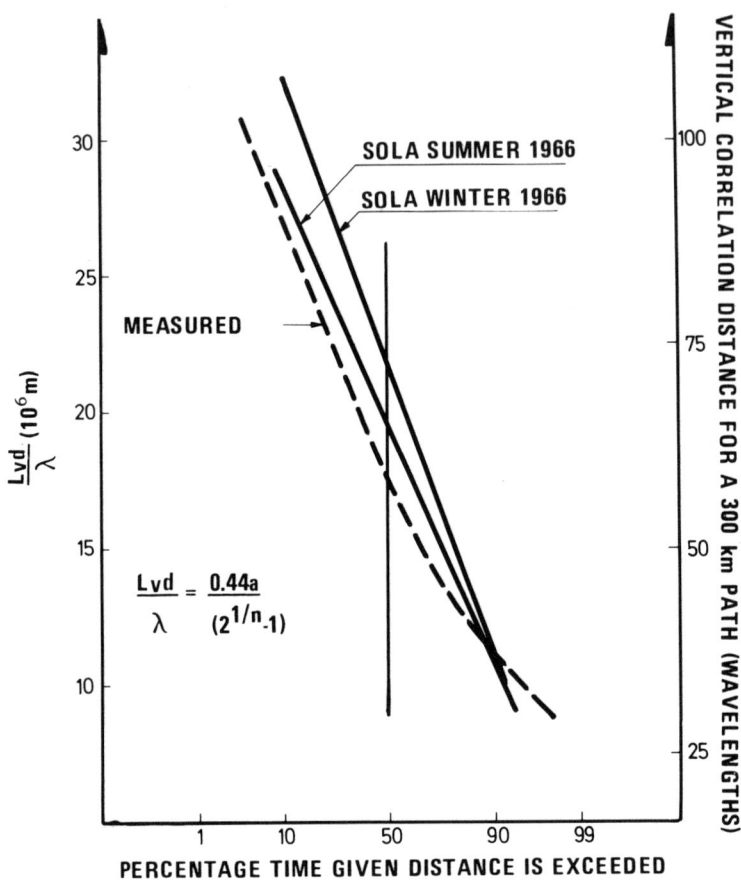

Figure 5.15 Vertical field strength correlation distance. Experimental results after Grosskopf [48].

The two limiting cases are an infinitesimally thin layer ($\Delta h = 0$) and an infinitely thick layer ($\Delta h = \infty$). For $\Delta h = 0$ it can readily be verified that the power reflection coefficient is given by

$$|\sigma|^2 = \left| \frac{\Delta n}{2 \sin^2 \theta/2} \right|^2 \tag{5.35}$$

i.e.,

$$P(\theta) \sim \theta^{-4} \tag{5.36}$$

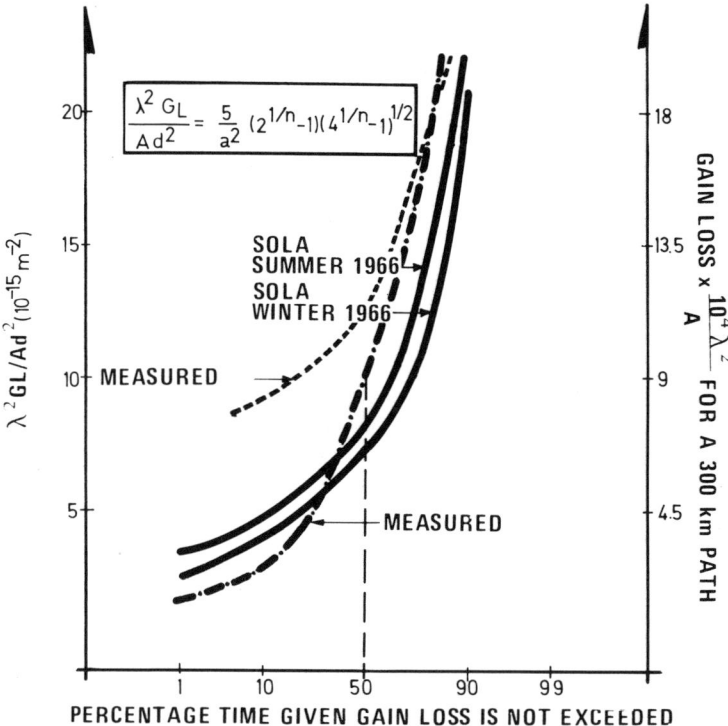

Figure 5.16 Antenna gain loss G_L. Experimental results after Grosskopf [48] and Hall [49].

We see that an infinitesimally thin layer (discontinuity in n) gives rise to a scattering-angle dependence of scattered power which is very close to the -11/3 law obtained for homogeneous isotropic inertial subrange turbulence. A layer of finite thickness results in an oscillatory dependence of the reflection coefficient on the scattering angle. The thicker the layer is, the more rapid are the oscillations. When the layer thickness approaches infinity, i.e., when we remove the upper knee of the profile, the oscillator angular dependence disappears since we no longer get interference between the two boundaries. The result is a smooth angular spectrum given by:

$$|\sigma|^2 = \left(\frac{dn}{dh} \frac{\lambda}{8\pi \sin^3 \theta/2}\right)^2 \qquad (5.37)$$

92 ADAPTIVE RADAR

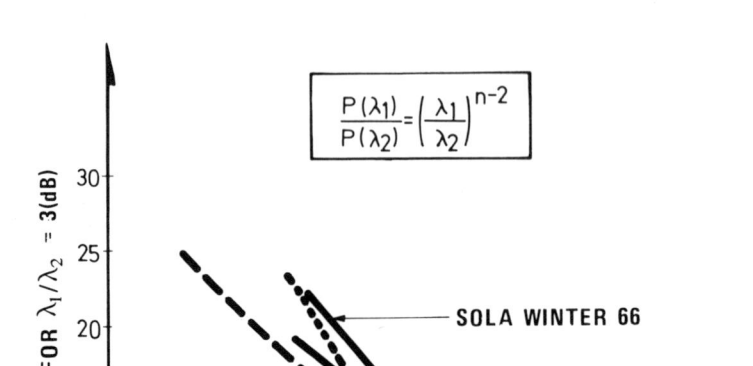

Figure 5.17 Ratio of received power for a wavelength ratio of 3. Experimental results after Eklund and Wickerts [39].

i.e., $P(\theta) \sim \theta^{-6}$. We may conclude, therefore, that a single atmospheric layer gives rise to angular dependences varying from θ^{-4} to θ^{-6}, depending on the layer thickness. The power scattered in a given direction is determined by layer thickness and by the change in refractive index through the layer.

However, no single layer can give an angular dependence weaker than that corresponding to the θ^{-4} law. Finally, we shall consider the spatially intermittent type structure shown on the

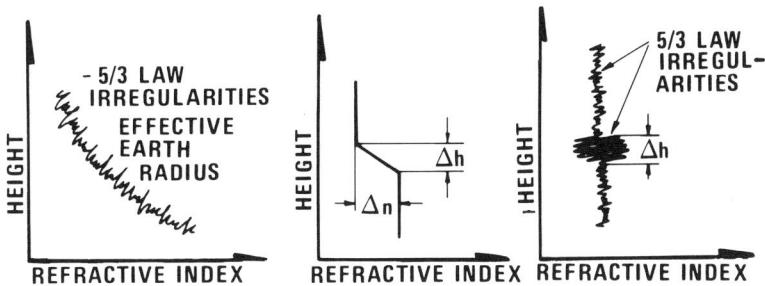

Figure 5.18 The categories of atmospheric structure under consideration.

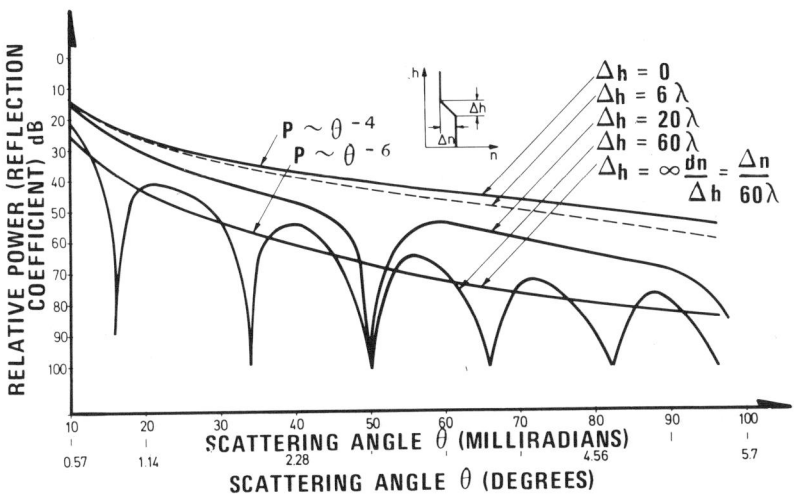

Figure 5.19 Angular power spectrum of a scattered wave resulting from a layer of thickness Δh.

right in Figure 5.18. We shall see to what extent such a structure influences the angular power spectrum.

To simplify the treatment, we shall assume that the variance of the refractivity irregularities through the turbulent stratum is given by a $(\sin x)/x$ relationship and that the spectrum of the irregularities can be written in the form K^{-n}.

The resultant spectrum is a convolution integral where the K^{-n} spectrum is convolved with the spectrum of the $(\sin x)/x$

94 ADAPTIVE RADAR

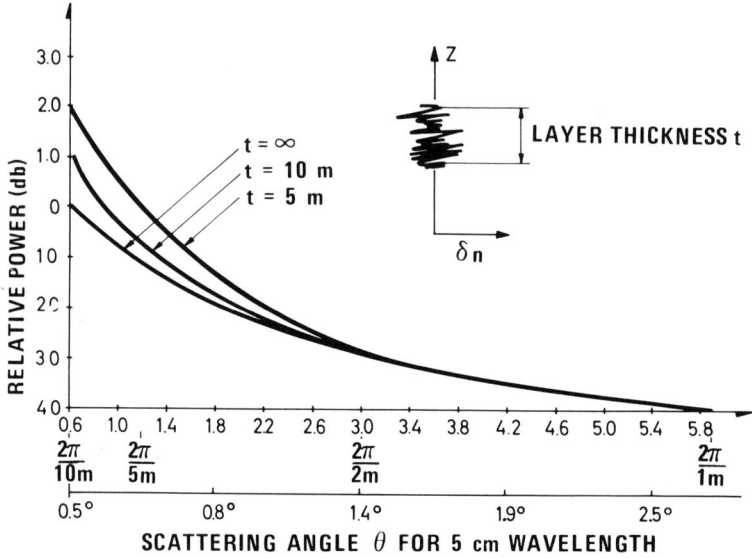

Figure 5.20 Angular power spectrum of a scattered wave resulting from a thin stratum of intense turbulence.

filter function. The spectrum of the (sin x)/x filter is a rectangular function. The width of this rectangular filter spectrum is taken to be $2K'$ and the density A is taken as $\frac{1}{2}K'$ such that the integral of the filter spectrum becomes unity. The power spectrum of such a turbulent stratum then becomes:

$$E(K) = \int_{K-K'}^{K+K'} ACK^{-n} \, dK$$

Solving this simple integral we get:

$$E(K) = AC \left\{ 2K'K^{-n} + \frac{2}{3!}(1-n)(-n-1) K'^3 K^{-n-2} + \frac{2}{5!}[(1-n)(-n-1)(-n-2)(-n-3) K'^5 K^{-n-4}] \right\} +$$

higher-order terms

Choosing then n = 4 ≈ 11/3, we get:

$$E(K) = K^{-4} [1 + 10/3 \, (K/K')^{-2} + 7 \, (K/K')^{-4} + \ldots] \quad (5.38)$$

Figure 5.20 shows a graphical representation of this equation. We see that unless the thickness of the stratum is less than 2 or 3 times the projected wavelength $\lambda/(\sin \theta/2)$, the spatial intermittency of the refractive index irregularities has no influence on the angular spectrum of the scattered wave.

Furthermore, we know that all irregularities within the scattering volume contribute to the scattered field strength at the receiver. A very thin turbulent layer must therefore have a very large variance in order to give a contribution which is comparable with that of the background turbulence in which the stratum is assumed to be embedded [50].

Finally, we see from Figure 5.20 that the existence of one turbulent layer results in a spectrum slope which is steeper than that one would have with −5/3 law homogeneous turbulence.

Transmission Loss as a Function of Carrier Frequency

Consider now the experiment involving simultaneous transmission and reception on two widely separated frequencies and scaled antennas. If the antenna beams are narrow, such that the scattering volume is determined by beam geometry and not by the scattering mechanism, the scattering volumes for the two frequencies are identical. Measuring the ratio of the power received on the two frequencies, we get information about the refractive index spectrum along a vertical direction within the scattering volume. Writing this spectrum as $\Phi(K) \sim K^{-n}$, the power ratio is simply given by:

$$\frac{P(\lambda_1)}{P(\lambda_2)} = (\lambda_1/\lambda_2)^{n-2} \qquad (5.39)$$

If the received power is normalized with respect to the free-space transmission loss, the corresponding power ratio becomes:

$$\frac{\dfrac{P}{P_{FS}}(\lambda_1)}{\dfrac{P}{P_{FS}}(\lambda_2)} = (\lambda_1/\lambda_2)^{n-4} \qquad (5.40)$$

We see that atmospheric refraction does not enter the equation in this case.

Spatial Correlation Properties of Field Strength

We have a wide-beam transmitter radiating its power essentially in a horizontal direction. The resulting scattered wave is received by two nearly identical, small-aperture receiving antennas positioned beyond the horizon relative to the transmitter.

The antennas are spaced vertically or horizontally such that the center line through the receiving antennas is normal to the line through the transmitter and the receivers. We measure the normalized complex correlation of the voltages induced in the antennas.

As we have already seen (Chapter 2), this spatial field-strength correlation function is the Fourier transform of the angular power spectrum of the wave reaching the receiving antennas. Thus if the antennas are spaced vertically, the correaltion distance of field-strength gives us information about the angle-of-arrival spectrum in the vertical plane. Similarly, horizontally spaced antennas give us information about the angular spectrum in a horizontal plane.

Referring to discussion above, the vertical correlation distance is given by:

$$\frac{L_V}{\lambda} = 0.44 \; \frac{a}{d} \; (2^{1/n} - 1)^{-1} \qquad (5.41)$$

and the horizontal correlation distance:

$$\frac{L_H}{\lambda} = 0.44 \; \frac{a}{d} \; (4^{1/n} - 1)^{-1/2} \qquad (5.42)$$

where a = effective Earth radius (i.e., refraction effects are important)
d = path length
n = slope of the irregularity spectrum within the scattering volume ($\Phi(K) \sim K^{-n}$)

The vertical correlation distance is determined by the vertical component of the three-dimensional spectrum, whereas the horizontal correlation distance is governed by the spectrum along a direction which is determined by the length of the path. For a line-of-sight path (if a scatter experiment could be realized on such a path), it is the horizontal component of the three-dimensional spectrum that matters. If the path length is large, it is the component along a direction which is close to being vertical that has the dominating influence [45].

Antenna Gain Degradation

As we have already seen the gain loss is inversely proportional to the product of the vertical and horizontal correlation distances. Thus, irrespective of whether we are dealing with a homogeneous random turbulent atmosphere or one which is characterized by a layered structure, the gain loss is given by:

$$G_L = \frac{A}{L_v L_H} = \frac{5(2^{1/n} - 1)(4^{1/n} - 1)^{1/2} A}{(a/d)^2 \lambda^2} \quad (5.43)$$

Note that this equation is based on the assumption that the transmitting antenna beam is broad both in azimuth and elevation and the receiving beam is narrow in both planes. This condition is necessary for the influence of the atmospheric structure on the gain loss to be a maximum.

Time Delay

The mechanisms we have discussed so far, all rely directly or indirectly on the angular power spectrum of the scattered wave. We have seen that the two extreme structure categories, homogeneous -11/3 law turbulence and single, infinitesimally thin layer, give angular spectra which for all practical purposes are indistinguishable.

Admittedly there are special sets of, for example, beamswinging experiments, that at least in principle can reveal the differences, but on the basis of simple single experiments and reasonable beamwidths, it is very difficult to analyze a single-layer structure.

The two remaining circuit parameters to be discussed, time delay and bandwidth measurements, do not rely on the angular power spectrum but on the delay function. Using wide-beam antennas on either end such that the multipath transmission is governed by the scattering mechanism rather than by beam geometry, we seek an expression relating path length ℓ and the position in space of the scattering element (layer). Note that the loci of constant delay in a forward scatter circuit are concentric elliposoidal surfaces having the transmitting and receiving points as foci. Discussing layers that are predominantly horizontal, we see that those in the neighborhood of the midpath point of interest in a forward-scatter experiment coincide with the constant-delay surface. It is thus essentially the vertical distribution of scatterers which matters in a pulse-delay experiment.

Simple geometry shows that if θ is the scattering angle corresponding to a given path length ℓ, and d is the length of the chord between the transmitter and the receiver, ℓ and θ are related by the equation:

$$\theta = 2\left[\left(\frac{\ell}{d}\right)^2 - 1\right]^{1/2} \tag{5.44}$$

Differentiating this equation to get an expression for $d\ell/d\theta$, we find the following relationship between $\Delta\theta$ and $\Delta\ell$:

$$\Delta\theta = \frac{4\ell}{d^2\theta} \Delta\ell \tag{5.45}$$

Neglecting second-order terms and putting $\Delta\theta = \Delta z/(d/2)$ and $\theta = d/a$, where Δz is the height coordinate measured from the intersecting point of the two tangent planes through T and R, and a, as usual, is the effective earth radius, we get the following expression:

$$\Delta\ell \approx \frac{2d}{a} \Delta z \tag{5.46}$$

If the refractive index irregularities are limited to a layer or thin stratum of thickness Δz, a delta pulse transmitted at T will appear as a broadened pulse, or a spectrum of delta pulses at R, and the width of this delay spectrum is given by:

$$\Delta\gamma = \frac{\Delta\ell}{c}$$

$$= \frac{2d}{ac} \Delta z \tag{5.47}$$

where c = the wave velocity

As an example, consider a 200-km path and a thickness of the turbulent stratum of 10 m. From Equation 5.46 we see that path length difference $\Delta \ell$ will be:

$$\Delta \ell = \frac{1}{15} \Delta z$$

The width of the delay spectrum $\Delta \gamma$ associated with a 10-m thick stratum will be:

$$\Delta \gamma \approx 2 \times 10^{-3} \, \mu sec$$

Assume now, that the thin layer is embedded in a background of turbulence and that this turbulence gives rise to a -11/3 law refractive index irregularity spectrum. We have previously shown that the 1/e width of the delay spectrum resulting from a homogeneous irregularity spectrum is given by:

$$\Delta \gamma = \frac{d^3}{8a^2 c} (e^{2/n} - 1) \qquad (5.48)$$

where n is the slope of the irregularity spectrum. This should be 11/3 for homogeneous isotropic inertial subrange turbulence. Using the same path geometry as in the above example, we find from Equation 5.48 that the width of the delay spectrum resulting from the background turbulence is:

$$\Delta \gamma = 60 \times 10^{-3} \, \mu sec$$

This is a factor of 30 relative to the delay spectrum width resulting from a 10-m thick layer. The background turbulence, however, will give a particular decrease in intensity with increasing delay whereas a thin layer of large refractive index variance will give a well defined peak in the delay spectrum.

Bandwidth

Our interest is now finally focused on the bandwidth properties of a scattered wave in relation to the particular atmospheric structures under consideration.

We have previously shown that there is a simple relationship between the width of the delay spectrum and the bandwidth. If $\Delta\gamma$ is the width of the rectangular delay spectrum resulting from a layer of thickness Δz, the bandwidth Δf is given by:

$$\Delta f = \frac{1}{\Delta\gamma} = \frac{c}{\Delta\ell}$$

From Equation 5.47 we get:

$$\Delta f = \frac{ac}{2d\Delta z}$$

On the basis of the previous example with a 10-m thick layer within the scattering volume limited by broad beams and a 200-km path, we find that the bandwidth is:

$$\Delta f = 500 \text{ MHz}$$

Then let us consider the bandwidth limitation resulting from the background turbulence. We have already shown that this bandwidth is given by:

$$\Delta f = \frac{8a^2 c}{2\pi d^3 (e^{4/n} - 1)} \tag{5.49}$$

For $n = 11/3$ (inertial subrange turbulence) we find that the bandwidth corresponding to a 200-km path is:

$$\Delta F = 1.2 \text{ MHz}$$

We see that the contribution to the bandwidth of the background turbulence is very small compared with that of a thin layer [51-58].

Table 5.2 summarizes the theoretically obtained relationships pertaining to forward scattering from turbulence and from stratified layers.

Table 5.2 Some Theoretical Relationships Pertaining to Forward Scattering from Turbulence and Stratified Layers

	Atmospheric Structure	
Propagation Parameter	Background Turbulence	Thin Layer
Angular Power Spectrum	$\frac{P(\theta)}{P(\theta_0)} = (1 + \frac{\theta}{d/a})^{-(11/3+1)}$	$\frac{P(\theta)}{P(\theta_0)} = (1 + \frac{\theta}{d/a})^{-4}$
Wavelength Dependence	$\frac{P(\lambda_1)}{P(\lambda_2)} = (\frac{\lambda_1}{\lambda_2})^{11/3-2}$	$\frac{P(\lambda_1)}{P(\lambda_2)} = (\frac{\lambda_1}{\lambda_2})^{4-2}$
Vertical E-correlation	$\frac{L_V}{\lambda} = \frac{0.44\, a}{d(2^{3/11} - 1)}$	$\frac{L_V}{\lambda} = \frac{0.44\, a}{d(2^{1/4} - 1)}$
Horizontal E-correlation	$\frac{L_H}{\lambda} = \frac{0.44\, a}{d(4^{3/11} - 1)^{\frac{1}{2}}}$	$\frac{L_V}{\lambda} = \frac{0.44\, a}{d(4^{1/4} - 1)^{\frac{1}{2}}}$
Coupling Loss	$G_L = \frac{5A(2^{3/11} - 1)(4^{3/11} - 1)^{\frac{1}{2}}}{(a/d)^2 \lambda^2}$	$G_L = \frac{5A(2^{1/4} - 1)(4^{1/4} - 1)^{\frac{1}{2}}}{(a/d)^2 \lambda^2}$
Pulse Delay	$\Delta\tau = \frac{d^3}{8a^2 c}(e^{6/11} - 1)$ 200-km path $\Delta\tau = 60$ nsec	$\Delta\tau = \frac{2d}{ac}\Delta h$ 200-km path, $h = 10$ m $\Delta\tau = 2$ nsec
Bandwidth	$\Delta f = \frac{8a^2 c}{2\pi d^3 (e^{12/11} - 1)}$ 200-km path $\Delta f = 1.2$ MHz	$\Delta f = \frac{ac}{2dh}$ 200-km path, $h = 10$ m $\Delta f = 500$ MHz

In summing up this section on propagation mechanisms involving scattering and reflection, the following should be noted: the parameters characterizing the transmission medium exhibit dramatic variations with time. When illuminating a target through this variable propagation medium, one should bear in mind that the propagation conditions can be favorable during small time intervals only. To achieve maximum radar target resolution, therefore, we shall have to take many "snapshots" and select the data sets which give optimum resolution (contrast).

Scattering by Particles (Rainfall)

A review on propagation mechanisms in relation to adaptive radar systems would be incomplete if the phenomena related to rainfall were not discussed. In this presentation a simple qualitative discussion will be given. For details, the reader is referred to the literature [4,59-61].

When studying the effect of precipitation on electromagnetic waves, several factors describing the rain structure should be considered. These are:

- rainfall rate,
- drop-size distribution,
- shape of raindrops, and
- canting angle of raindrops (orientation in space of ellipsoidal raindrops).

Rainfall rate and drop size distribution are, evidently, the most important parameters.

Consider a dielectric sphere of diameter D in an electromagnetic field where the wavelength λ is large compared with the diameter of the sphere. The sphere will give rise to an induced dipole moment as discussed in Chapter 2, and reradiate the power which has been extracted from the incident electromagnetic field in all directions.

The region where the diameter of the sphere is very large compared with the wavelength is known as the Rayleigh region. The analysis related to this case is well known and leads to the result that the scattering cross section of the sphere increases as the fourth power of frequency. If the sphere has a complex dielectric constant, part of the power incident on the sphere is dissipated as heat and, consequently, causes a reduction in the amount of reradiated power.

If the frequency of incident radiation is increased such that its wavelength becomes comparable with the dimensions of the dielectric sphere, the problem rapidly becomes complex. One now will have to consider the field patterns within the sphere. The simplest pattern is that obtained when the circumference of the sphere is one wavelength. We then have a condition which is referred to as dipole resonance giving maximum scattering cross section. Increasing the frequency further, a situation characterized by quadrupole resonance occurs when the circumference is two wavelengths long. Decreasing the wavelength of the field still further, hexapole resonance conditions are reached, followed by octapole, and so on until the cross section becomes a highly complex linear superposition of these multipole radiators which finally converge to the limit known as the "geometrical" value.

Normalizing the scattering cross section to this value, the scattering cross section vs frequency dependence is shown in Figure 5.21.

Diffraction of Radio Waves by Obstacles

Radio propagation over natural obstructions has been a subject of considerable interest. Classical knife-edge diffraction theory is used as an idealization of the physical problem. However, due to difficulties in obtaining an analytical solution, particularly when ground reflections and complex obstacles are considered, most of the previous work has been restricted to compilations of results from theory and measurements, i.e., magnitude considerations only.

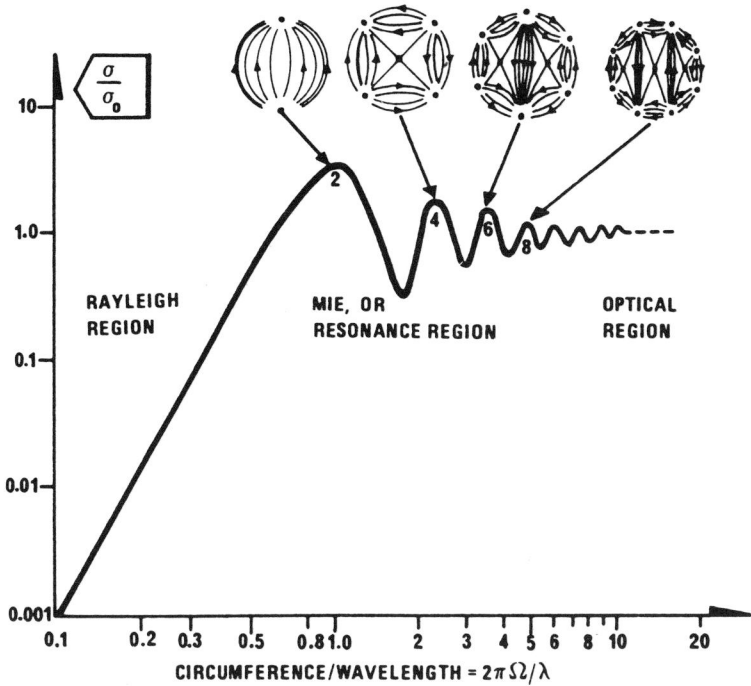

Figure 5.21 Normalized scattering cross section of a dielectric sphere as a function of normalized wavelength.

The knife-edge diffraction problem has been solved formally by several techniques. The early work by Sommerfeld, who based his analysis on the wave equation with appropriate boundary conditions, is perfectly rigorous and not restricted to small diffracting angles, but generalization to the multiple-reflection problem is not readily accomplished.

Fresnel and Kirchhoff based their work on an analytic formulation of Huygen's principle. The Fresnel-Kirchhoff method has the advantage of the simplicity of ray-path concept, and is used almost exclusively in radio propagation literature. However, the method suffers from lack of rigor and from the appearance of the Fresnel integrals which tend to obscure physical interpretations. To obtain practical solutions to the problem of diffraction in the presence of multiple reflections, a

106 ADAPTIVE RADAR

phasor summation of the Fresnel-Kirchhoff results is performed for the single ray problem.

In a paper by Ratcliffe [62] a very neat and general method for solving complex diffraction problems is sketched. Ratcliffe shows that if, in particular, a one-dimensional diffracting screen is illuminated by a wave (wavelength λ), the angular power spectrum $P(\theta)$ produced, expressed in terms of $\sin \theta$, is the Fourier transform of the spatial distribution $E(z)$ of the wave front just as it leaves the screen, provided z is expressed in terms of λ.

Basic Theory of Diffraction

Our problem is as shown in Figure 5.22. The transmitter illuminates a region above the obstacle. This gives rise to a distribution of field strength along a vertical direction above the obstacle. This distribution is determined by multipath effects, the ground reflection coefficient and the radiation properties of the antenna. On the basis of knowledge about $E(z)$, we want to calculate the angular power spectrum $P(\theta)$ of the diffracted wave.

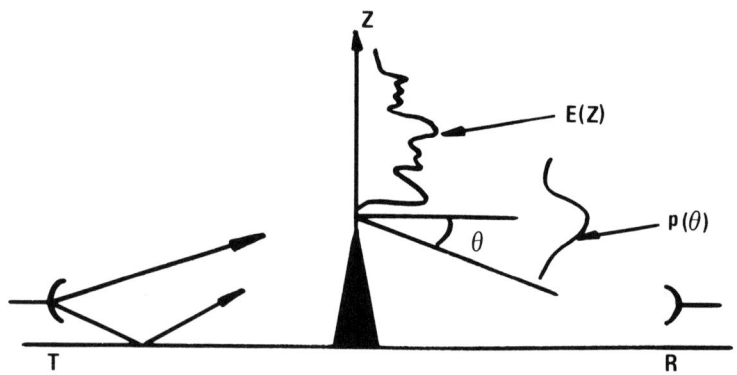

Figure 5.22 Geometry of the diffraction problem.

We now use exactly the same approach as for the scatter propagation (case discussed in Chapter 2) and derive the following expression for the secondary field E_s (scattered field or diffracted field, as the case may be):

$$E_s(K) = \frac{k^2}{4\pi R} \int E_0(z) \epsilon(z) e^{-jKz} dz \qquad (5.50)$$

where R = distance between receiver and diffraction element (distance from obstacle)
$|k|$ = $2\pi/\lambda$ (λ is the wavelength)
$\vec{K}$ = $\vec{k}_{in} - \vec{k}_{sc}$ $|K| = 4\pi/\lambda \sin\theta/2$
$\epsilon(z)$ = distribution of permittivity above the obstacle
$E_0(z)$ = distribution of field strength above the obstacle
θ = essentially the direction of the diffracted wave under consideration.

Thus, the angular distribution ($\vec{K}$ distribution) of the secondary field is the Fourier transform of the $E_0(z) \epsilon(z)$ product.

Two limiting cases are of interest:

1. If the field E_0 incident on the plane above the obstacle varies far more rapidly with height z than $\epsilon(z)$ does, $\epsilon(z)$ does not contribute to the convolution, and we can put $\epsilon(z)$ outside the Fourier integral. We thus have an important relation: the angular distribution of the diffracted wave is the Fourier transform of the field strength distribution measured in a plane above the obstacle giving rise to diffraction.
2. Conversely, if $\epsilon(z)$ varies rapidly in relation to $E_0(z)$, $E_0(z)$ does not contribute to the convolution and we have an expression for the scattered wave. The angular distribution of the scattered waves is the Fourier transform of the permittivity (refractive index) distribution.

Thus, if we want to calculate the transmission loss resulting from a given obstacle, the procedure is in short the following:

Step 1

Calculate the field-strength distribution along a vertical direction over the obstacle, considering the influence of ground reflections, etc. Note that if the obstacle cannot be considered as a knife edge, then reflection from the obstacle itself may contribute to the field-strength distribution over the ridge.

Step 2

Having obtained an expression for $E_o(z)$, we compute the Fourier transform of $E_o(z)$, thus obtaining the angular field-strength distribution $E_s(\theta)$ of the diffracted wave. To obtain the angular power distribution $P(\theta)$, we shall have to multiply $E_s(\theta)$ with its complex conjugate $E_s^*(\theta)$.

Note that for this Fourier transformation to be dimensionally meaningful, the space coordinate will have to be normalized with respect to wavelength. The direction θ is then to be expressed in terms of $\sin \theta$.

Step 3

Having obtained the angular power spectrum $P(\theta)$, we shall essentially have to repeat step 1 to obtain the desired expression for the power received at a given point in the diffraction zone behind the obstacle.

We shall now consider this procedure in some degree of detail. Let us then first calculate the field strength distribution above the obstacle. To obtain simple, comparatively general and (particularly) physically interpretable expressions for the field-strength distribution above the obstacle (step 1), we shall have to idealize the problem and in doing so make certain approximations. Let us consider the case with one direct and one reflected wave. We then see that the most striking feature of the vertical field-strength distribution is the periodicity. We clearly have a component of field strength which varies in a sinusoidal fashion with height.

From knowledge about the reflection properties of surfaces under various degrees of roughness, one would expect the influence of multipath on the vertical field strength profile to diminish with height. Hence one would expect the sinusoidal field strength oscillation to be damped in some manner. Lastly, we see that the DC level of the $E_o(z)$ profile is a parameter of some importance.

Our diffraction problem will therefore be based on the following field strength profile (Figure 5.23):

- a constant term E_1, and
- a damped sinusoidal term given by $E_2(z) = E_2 \, e^{-\alpha z k} \sin 2\pi \, z/L$

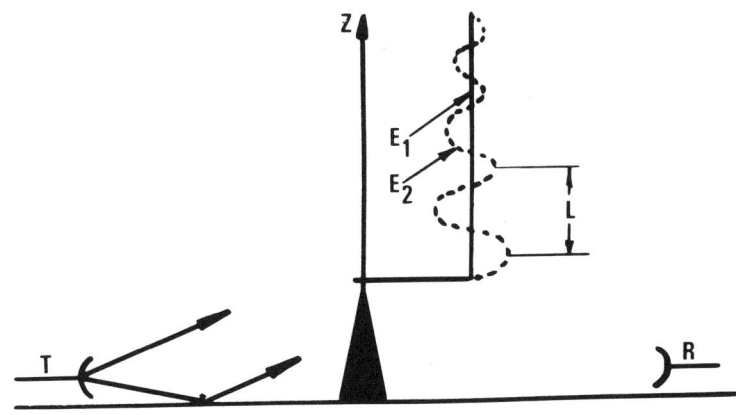

Figure 5.23 Vertical field-strength distribution with ground reflections.

110 ADAPTIVE RADAR

Angular Spectrum of Diffracted Field. The next step in the calculation involves a Fourier transformation of the vertical field strength profile.

The first term in the expression for the field is a step function:

$$E_1(z) = 0 \text{ for } z < 0$$
$$= 1 \text{ for } z > 0$$

The Fourier transform of this is:

$$F_1(\theta) = \tfrac{1}{2}\delta(\theta) + \frac{1}{j\theta} \qquad (5.51)$$

This assumes small angles θ such that $\theta \approx \sin\theta$. Note that δ is a Dirac delta function.

For simplicity we write the second term as:

$$E_2(z) = E_2(e^{-\alpha z k}\sin\beta z k) \qquad (5.52)$$

such that $\beta = \lambda/L$. This function gives a comparatively simple Fourier transform:

$$F_2(\theta) = \frac{\beta}{\alpha^2 + \beta^2 - \theta^2 + 2j\alpha\theta} \qquad (5.53)$$

such that the resultant angular spectrum is given by:

$$F(\theta) = F_2(\theta) + F_1(\theta)$$
$$= \frac{\beta E_2}{\alpha^2 + \beta^2 - \theta^2 + 2j\alpha\theta} + \frac{E_1}{j\theta} \qquad (5.54)$$

The angular power spectrum is then obtained from:

$$P(\theta) = F(\theta) \cdot F^*(\theta)$$

This function has a pronounced maximum for a certain direction, namely the direction corresponding to:

$$\theta = \pm(\beta^2 + \alpha^2)^{1/2}$$

If we have no damping, i.e., if $\alpha = 0$, we have a Dirac delta function for $F_2(\theta)$ centered at:

$$\theta = \beta \text{ i e at } \theta = \frac{\lambda}{L}$$

This is the same as the results obtained using the Bragg scattering relationship in connection with the scattering from sinusoidal variations with height of the refractive index. From Bragg we have the following. Maximum scatter in a direction θ given by:

$$K = \frac{4\pi}{\lambda} \sin \theta/2$$

where $K = 2\pi/L$. As in the diffraction case above, L is the period in space of the sinusoidal variations, i.e.,

$$\frac{2\pi}{L} = \frac{4\pi}{\lambda} \theta/2$$

i.e., $\theta = \lambda/L$ as we obtained above. We know that a single ground-reflected component interfering with the direct wave will give a field strength distribution above the obstacle of the form given in Equation 5.54. The damping factor α will then be determined by the rate at which the ground reflection coefficient decreases with increasing angle of incidence to the ground. The period of oscillation, L, is given by the geometry. Let R be the distance between transmitter and knife-edge obstacle, h the height of the transmitter above the flat earth and λ the radio wavelength; then the period L is given by:

$$L = \frac{\lambda R}{2h} \qquad (5.55)$$

Thus, in this case, we will have maximum energy diffracted in a direction given by:

$$\theta = \frac{2h}{R} \qquad (5.56)$$

Thus, by adjusting the height, h, of the transmitting antenna, we can beam the diffracted wave in the desired direction. We see that ground reflection (leading to an oscillating vertical profile of field strength) actually increases the power received at a given point behind the obstacle. This increase is known as "obstacle gain," or "multipath gain." Note that we shall have to transform the coordinate system such that θ is measured relative to the line joining the top of the knife-edge obstacle and the transmitter and not relative to the horizontal line through the obstacle top.

Let us now, as an example, calculate the obstacle gain. As we have seen, the diffracted field F_1 resulting from the unit step function (the DC term) is given by:

$$F_1 = \frac{E_1}{j\theta}$$

The maximum field resulting from the oscillating term (i.e., field in direction θ where $\theta = (\alpha^2 + \beta^2)^{1/2}$ is, as seen from Equation 5.54, given by:

$$F_2 = \frac{E_2 \beta}{j 2\alpha (\alpha^2 + \beta^2)^{1/2}} \qquad (5.57)$$

Obstacle gain is then:

$$G_o = [(F_2 + F_1)/F_1]^2 = [(E_2/E_1)(\beta/2\alpha) + 1]^2 \qquad (5.58)$$

Note that this gain refers to the direction $\theta = (\alpha^2 + \beta^2)^{1/2} = [(\lambda/L)^2 + \alpha^2]^{1/2}$, where L is the spatial period of field strength oscillations.

We see from this equation that the diffraction loss relative to that of the no-multipath case can be reduced considerably by a proper adapting of the appropriate parameters:

1. making the field strength oscillations above the obstacle as deep as possible by making the direct wave comparable in magnitude with that of the interfering wave (in equation (5.54) $E_2 \to E_1$), and
2. reducing the damping factor α by seeking ground reflections from a surface giving a small decrease in reflection coefficient with angle of incidence.

Tables 5.3 and 5.4 give some practical examples. In the same way as the Tables 5.3 and 5.4 give the obstacle gain for various depths of the field strength oscillations and for various damping factors, Figures 5.24 and 5.25 show the whole spectrum in the diffraction angle range from 0 to some 20 m radians.

Table 5.3 Obstacle Gain as a Function of the Depth of the Field-Strength Oscillations above the Obstacle[a]

	E_1/E_2					
	1	0.8	0.6	0.4	0.2	0
Obstacle Gain G_o (dB)	10.8	9.5	7.9	6	3.5	0

[a] $\alpha/\beta = 0.2$

Table 5.4 Obstacle Gain as a Function of Damping Factor α[a]

	α/β				
	0.2	0.4	0.6	0.8	1.0
Obstacle Gain G_o (dB)	10.8	7.04	5.26	4.2	3.5

[a] $E_2/E_1 = 1$.

We see from Figure 5.24 that the higher the damping factor (for a given period $L = \lambda/\beta$ of the field strength oscillations), the lower is the obstacle gain.

Figure 5.25 shows the effect of varying the oscillation period L. We see that changing L results in a change in the direction of maximum diffraction whereas the obstacle gain remains constant since E_2/E_1 and a/β are kept constant.

We have shown that a terrain obstacle within line of sight gives rise to a well-defined beaming of the diffracted power behind the obstacle. The direction in which the power is beamed is determined by factors such as the height above ground of the transmitting antenna. The pathloss can be greatly reduced by applying general adaption schemes.

The theoretical approach on which the numerical results are based is very simple. Knowing the path and obstacle geometry and knowing the properties of the reflecting ground, the diffracted power spectrum can readily be optimized.

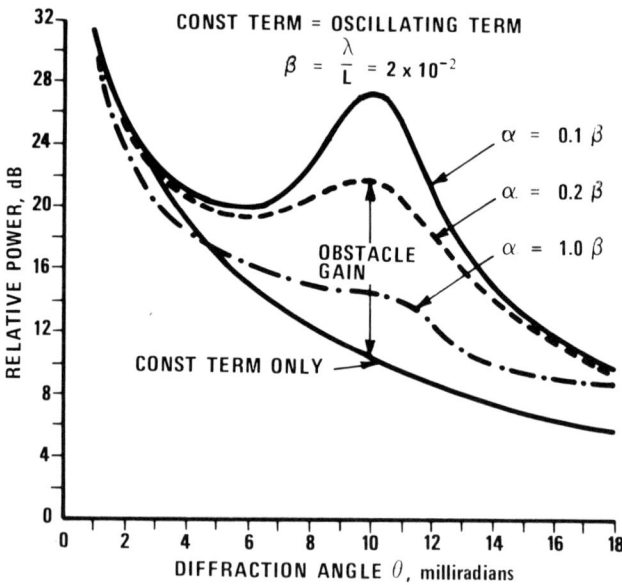

Figure 5.24 Angular diffraction spectrum for various damping factors α.

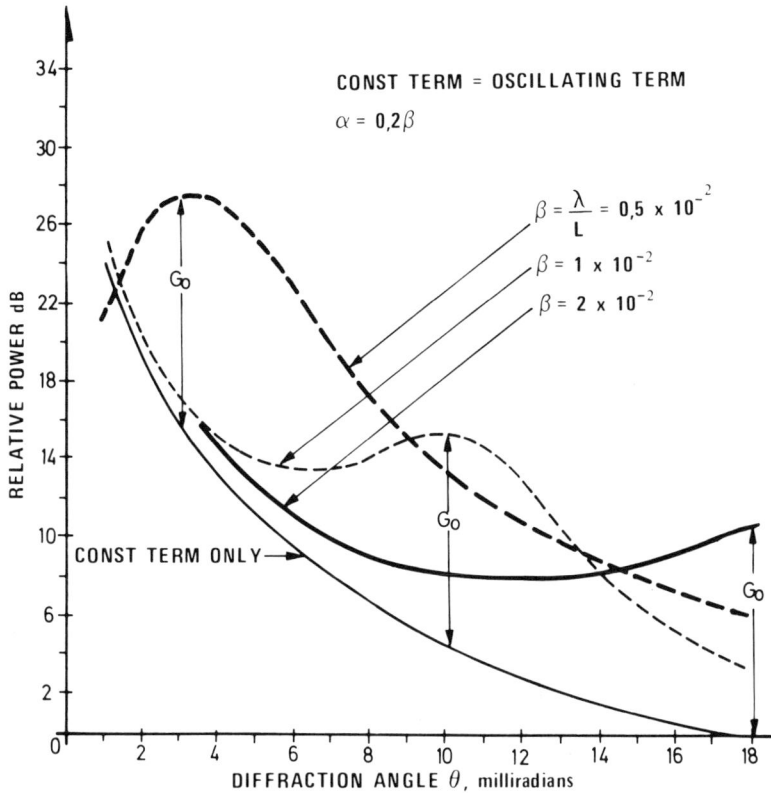

Figure 5.25 Angular diffraction spectrum for various periods of field-strength oscillations.

Bandwidth Limitations of a Transmission Path Involving Diffraction

Having formed the bases for calculating the path loss and obstacle gain, we shall now consider the bandwidth-limitations of a transmission path involving diffraction. As in the sections above, the aim is to provide a method which first of all is physically interpretable and which makes it possible to form an opinion as to the relative importance of the various factors involved. Secondly, the aim is to form the basis for approximate calculations.

116 ADAPTIVE RADAR

In Chapter 2 we saw that the bandwidth of a transmission circuit is directly determined by the delay function. Knowing the delay function, the bandwidth function (power spectrum) is obtained directly from the delay function by a simple Fourier transformation process. From Figure 5.26 we see that if the illuminated area above the obstacle is wide, the delay function will be correspondingly wide, and the bandwidth function will be narrow. We shall now consider the case where the illumination is limited to an antenna pattern of width β with elevation angle $\theta/2$ such that the total diffraction angle for a symmetrical path will be θ.

From simple geometry we see that the path length difference is given by:

$$\Delta \ell = d/2 \, (\theta \beta + \beta^2)$$

$$\theta = \frac{4H_o}{d}$$

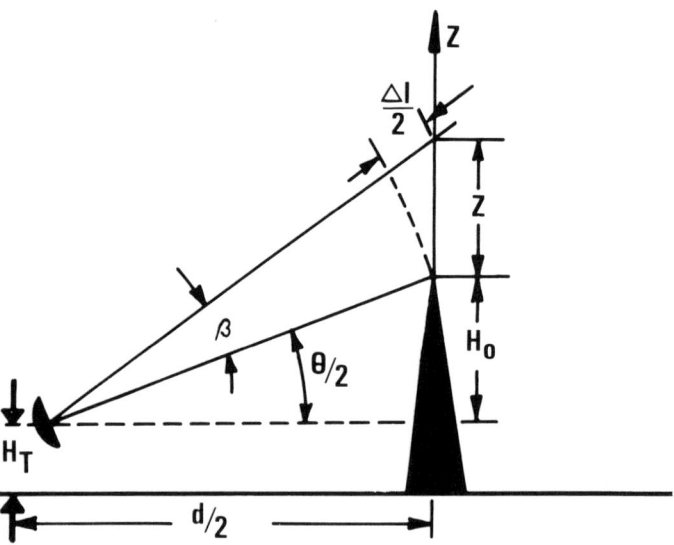

Figure 5.26 Simple path geometry for a transmission path involving knife-edge diffraction.

and

$$\beta = \frac{2z}{d}$$

$$\Delta \ell = \frac{2}{d}(2H_o z + z^2) \quad (5.59)$$

Having obtained the simple relationship between path length and ray geometry, let us return to Figure 5.23. Let us again assume that the field-strength distribution above the obstacle is characterized by a damped sinusoidal oscillation superimposed on an unit step function.

Let us denote the spatial period of the oscillations as δz and the damping function Δz. As we have seen in the section above, two limiting cases are of particular interest: first we assume that the ground reflections are such that the oscillating term (E_2 in Figure 5.23) dominates over the constant term E_1. In this case we essentially have a diffraction grating with line spacing δz. The object now is to calculate the bandwidth of such a diffraction circuit. As we have noted from the sections above, the center frequency in the bandpass filter is:

$$F_o = \frac{c}{\delta \ell}$$

where $\delta \ell$ is the increase in path delay associated with a vertical displacement δz. From Equation 5.59, we have already established the relationship between $\delta \ell$ and δz. Making use of this, we find that the center frequency of our filter is given by:

$$F_o = \frac{c}{\frac{2}{d}(2H_o \delta z + \delta z^2)} \quad (5.60)$$

118 ADAPTIVE RADAR

As we have seen above, we can calculate δz directly from the path geometry. If H_T be the height of the transmitting antenna above the reflecting plane, then:

$$\delta z = \frac{\lambda d}{4 H_T} \quad (5.61)$$

Making use of this relationship, the expression for the center frequency of our bandpass filter becomes:

$$F_o = \frac{c H_T}{H_o \lambda + \frac{\lambda^2 d}{8 H_T}} \quad (5.62)$$

Then let us calculate the width of this bandpass filter. As we have already seen (Equation 2.10 and Figure 2.3), if the delay function is an exponentially damped sinusoid with 1/e width equal to $\Delta \ell$, the half-power halfwidth of the resulting bandpass filter is equal to

$$\Delta F_{1/2} = 0.16 \frac{c}{\Delta \ell} \quad (5.63)$$

Accordingly, we find the bandwidth $\Delta F_{1/2}$ in the same way as we found the center frequency F_o above by substituting for $\Delta \ell$ using the appropriate value for Δz. If the geometry is such that:

$$2 H_o \delta z \gg \delta z^2$$

which requires:

$$H_o H_T \gg \frac{\lambda d}{8}$$

the expression for the center frequency F_o (Equation 5.62) deduces to:

$$F_o = \frac{c}{\lambda} \frac{H_T}{H_o} \qquad (5.64)$$

If we express the width of the exponential damping function Δz in terms of the oscillation period δz as:

$$\Delta z = n \cdot \delta z$$

our bandwidth becomes:

$$\Delta F_{1/2} = \frac{0.16}{n} \frac{c}{\lambda} \frac{H_T}{H_o} \qquad (5.65)$$

Note that the factor n is determined by factors such as the rate at which the reflection coefficient of the ground decreases with increasing angle of incidence. This is shown in Figure 5.27. Then consider the case with no ground reflections. We illuminate the area above the obstacle by an antenna beam the width of which is β. The width of the delay function is then directly given as:

$$\Delta \ell = \frac{d}{2} (\theta \beta + \beta^2)$$
$$= \frac{2}{d} (2H_o z + z^2) \qquad (5.66)$$

If we assume a gaussian distribution of radiated power through the antenna beam, the half-power half bandwidth is given by:

$$\Delta F_{1/2} = 0.37 \frac{c}{\Delta \ell} \qquad (5.67)$$

120 ADAPTIVE RADAR

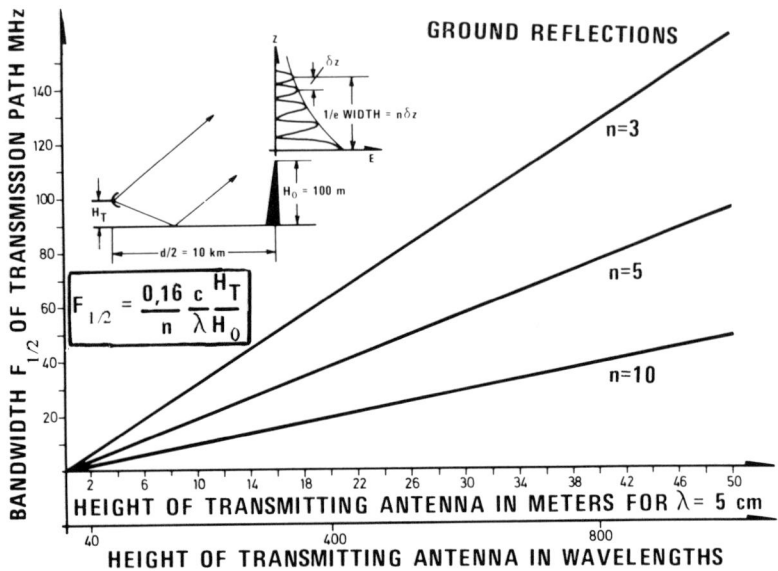

Figure 5.27 Bandwidth vs height of transmitting antenna for a diffraction path involving ground reflections. The example is for a path length of 20 km and obstacle height of 100 m.

Expressing now the vertical extent z of the illuminated area above the obstacle as $z = d/2 \cdot \beta$ the half-power beamwidth is given by:

$$\Delta F_{1/2} = \frac{0.37 \, c}{2H_o \beta + \frac{d}{2} \beta^2} \qquad (5.68)$$

This is shown in Figure 5.28.

This equation suggests that the bandwidth of the diffraction path diminishes as the beamwidth increases.

When we increase the beamwidth β, the situation arises where the "effective beamwidth" is determined by the diffraction process and not by beam geometry. With no ground reflections we have shown that the scattered field falls off inversely with diffraction angle θ.

RADIO PROPAGATION THROUGH THE ATMOSPHERE

Writing then the angular power spectrum of the diffracted wave as $P \sim \theta^{-2}$, we have:

$$\frac{P(\theta)}{P(\theta_o)} = \left(\frac{\theta_o + \Delta\theta}{\theta_o}\right)^{-2}$$

Then let us define the "effective beamwidth" as the angular region within which the diffracted power is larger than 1/10 of the maximum power obtained for the minimum diffraction angle $\theta_o = 2H_o/(d/2)$. We then have:

$$\frac{P(\theta)}{P(\theta_o)} = \frac{1}{10} = \left(\frac{\theta_o + 2\beta_{eff}}{\theta_o}\right)^{-2}$$

$$\beta_{eff} = \frac{\theta_o}{2}(\sqrt{10} - 1)$$

$$= \frac{2H_o}{d}(\sqrt{10} - 1)$$

Referring to the example illustrated in Figure 5.28, this means that as the beamwidth increases, the bandwidth will approach a limit given by the expression:

$$(\Delta F_{\frac{1}{2}})_{LIM} = \frac{0.37\,c}{2H_o\,\beta_{eff} + \frac{d}{2}\,\beta^2_{eff}}$$

Inserting the appropriate numbers for obstacle height H_o and range d, we find that the limiting bandwidth is 12 MHz.

If we are dealing with large antennas, such that β is small and if the height of the obstacle, H_o is large, the expression for bandwidth reduces to:

$$\Delta F_{\frac{1}{2}} = \frac{0.18\,c}{H_o \beta} \qquad (5.69)$$

122 ADAPTIVE RADAR

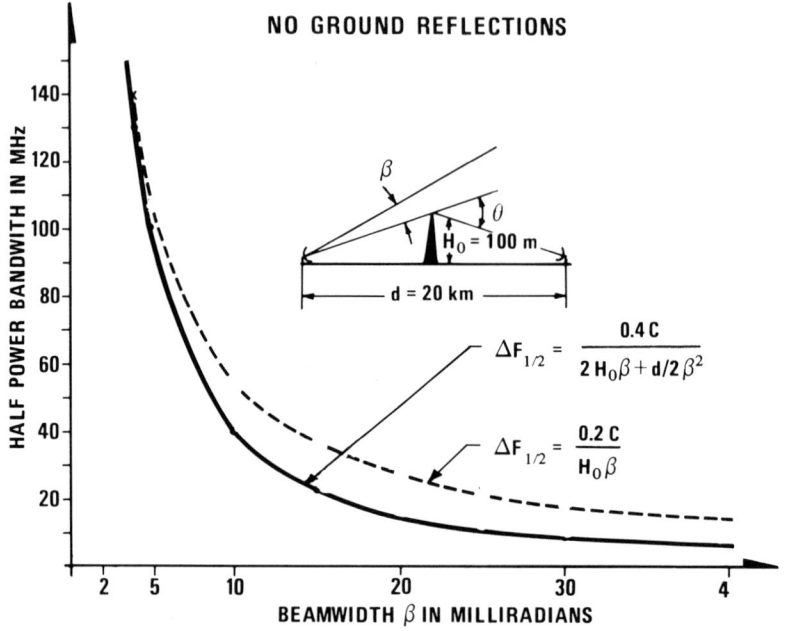

Figure 5.28 Bandwidth vs beamwidth for a diffraction path with no ground reflections. The example is for a pathlength of 20 km and obstacle height of 100 m.

Finally, Figure 5.29 summarizes the section on bandwidth properties of a transmission path involving knife-edge diffraction.

From the summarizing Figure 5.29 we see that if no ground reflections are involved, the transmission circuit will act as a lowpass filter (maximum at carrier frequency). If, however, ground reflections are involved, the field-strength distribution above the obstacle will be periodic giving rise to a bandpass filter which is not centered at the carrier frequency. This phenomenon is treated in some detail in Chapter 2, the results of which are summarized in Figure 2.3. Here we see that a periodic delay function results in a frequency covariance function which is displaced from the carrier frequency.

From the two preceding sections, we have shown that it is possible to minimize the transmission loss by adjusting the height of the transmitting antenna (measured in wavelengths) to "beam" the diffracted wave in the desired direction. We have also shown that we can adjust the geometry to optimize the

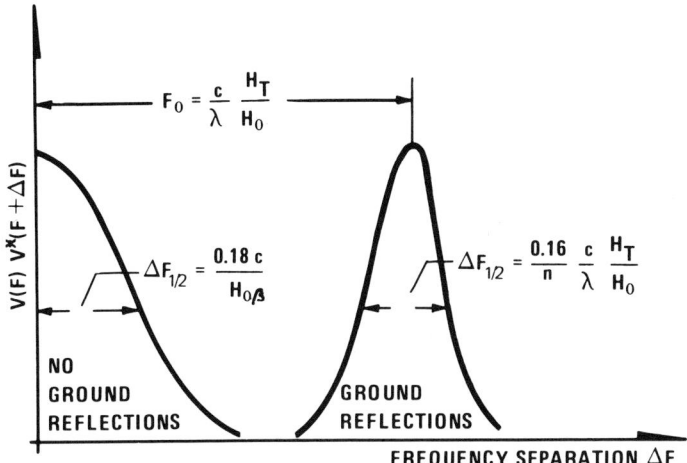

Figure 5.29 Bandwidth properties of a transmission path involving knife-edge diffraction.

bandwidth of the transmission circuit. It remains to study the spatial correlation properties of the scattered field. In Chapter 2 we showed that if we are to analyze an object, we shall have to produce an illuminating field which is coherent in phase and amplitude across the object. This means that if the height of the target to be analyzed is ΔH, we shall have to organize ourselves so as to produce an illuminating field which is coherent over the vertical region ΔH.

We shall now calculate the vertical correlation properties of field strength at the receiving site of a transmission path involving knife-edge diffraction. We shall base these calculations on the conclusions of earlier sections. Based on these earlier findings, we can derive accurate expressions if we base our calculations on regular Fourier transforms.

We shall now apply an approximate method based on semi-intuitive arguments to ensure a detailed physical understanding of the principles involved.

Spatial Correlation Properties of Diffracted Field

From Figure 5.30 we see that a transmitting antenna produces a periodic field strength distribution above the obstacle

124 ADAPTIVE RADAR

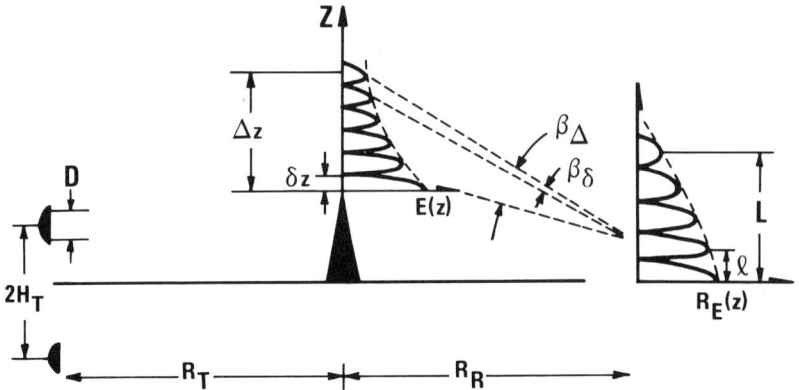

Figure 5.30 Spatial correlation of field-strength at the receiving site of a transmission path involving knife-edge diffraction.

when ground reflections are involved. If the ground acts as a perfect mirror, the transmitter can be visualized as having two antennas with vertical spacing H_T giving rise to a set of antenna lobes of beamwidth $\lambda/2H_T$. If we are dealing with antennas of diameter D, these antenna lobes will not be of the same intensity, but have the appearance of damped oscillations. The width of this damping function is determined by the antenna aperture such that the width of the envelope will be λ/D. Hence:

$$\delta z = \frac{\lambda}{2H_T} R_T$$

and

$$\Delta z = \frac{\lambda}{D} R_T$$

Having obtained qualitative information about the field-strength distribution above the obstacle, let us now discuss what this will lead to at the receiving site. The shortest scale to be observed in the vertical field-strength distribution at the receiver is associated with the widest angular distribution (as we have seen from Chapter 2). Conversely, the largest scale is associated with a narrow angular distribution.

Viewing the diffracting screen from the receiving end, we will observe an angle of arrival-spectrum with many lobes. As depicted in Figure 5.30, the overall width of the angular spectrum is therefore:

$$\beta_\Delta = \frac{\Delta z}{R_R} = \frac{\lambda R_T}{D R_R} \qquad (5.70)$$

and similarly

$$\beta_\delta = \frac{\delta z}{R_R} = \frac{\lambda R_T}{2 H_T R_R} \qquad (5.71)$$

We have already seen that an antenna with aperture D produces a beam of width

$$\beta = \frac{\lambda}{D}$$

By pure reciprocity considerations, therefore, an angle of arrival spectrum of width β will produce a field-strength distribution the width of which is

$$L = \frac{\lambda}{\beta}$$

We know, of course, that this is only approximately correct. If we want an accurate expression, the angular power spectrum is obtained as the Fourier transform of the field-strength distribution over the aperture. In the same way as the spatial correlation of field strength is obtained by Fourier-transforming the angular power spectrum. Making use of these simple qualitative results, we find that the fine-scale spatial correlation distance of the diffraction field is:

$$\ell = \frac{\lambda}{\beta_\Delta} = D \frac{R_R}{R_T} \qquad (5.72)$$

similarly, the large-scale correlation distance is:

$$L_1 = \frac{\lambda}{\beta_\delta} = 2 H_T \frac{R_R}{R_T} \qquad (5.73)$$

Intuitively, this is rather obvious: for a symmetrical path ($R_R = R_T$) the fine-scale field-strength structure is the same as the aperture of the transmitting antenna, whereas the large-scale structure is equal to twice the antenna height, which of course is the equivalent antenna aperture of a transmitting system involving ground reflections.

If we have no ground reflections, the correlation distance of field strength at the receiving sight will be:

$$\ell = D \frac{R_R}{R_T} \qquad (5.74)$$

In conclusion, we should note that the spatial correlation properties of the diffracted field are determined entirely by the path geometry, and the size and positioning of the transmitting antenna.

ABSORPTION PHENOMENA

So far we have concerned ourselves with the influence of the atmosphere in regard to spatial resolution of an adaptive multifrequency radar system. We have been focusing the attention of scattering and diffraction phenomena giving rise to contrast loss, to blurring.

We have this far neglected the effect of atmospheric absorption and assumed that the free-space attenuation loss is governed by the inverse square law. We shall now consider the role of absorption phenomena.

Absorption by Gases

To understand fully the mechanisms giving rise to absorption of radio waves, it is necessary to go into some detail and consider the microstructure of the gas. Let us start with the atom.

Atomic Spectra

In the range 1–500 GHz now covered by microwave sources,

direct transitions between fine or hyperfine levels of many atoms are theoretically observable. Because of the low intensity of the microwave transitions of atoms, these atomic absorption spectra are of very little practical importance in connection with atmospheric absorption. Then let us consider the next degree of smallness, the molecule.

Molecular Spectra

There are four different mechanisms giving rise to molecular absorption spectra.

Pure Rotational Spectra. This category constitutes the principal class of microwave molecular spectra. The spectra arise from the transitions between quantized rotational energies of molecules in single electronic ground states. If the molecule is excited by an electromagnetic field of a frequency ν corresponding to the energy difference between the ground state and the next energy level ($h\nu = E_1 - E_2$), the molecule will absorb this energy difference from the electromagnetic field resulting in an absorption line at this very frequency. The only atmospheric gases offering absorption through the mechanism under consideration are SO_2, O_3 and H_2O.

Inversion Spectra. Inversion may be described as the reflection of all the nuclei of a nonplanar molecule at its center of mass. In this way a new equilibrium configuration of the molecule is obtained which in nonplanar molecules cannot be obtained from the original one through any succession of simple rotations. There is no important atmospheric gas possessing inversion spectra.

Vibrational Spectra. Consider a molecule within which potential forces are acting; the atoms are thus bound together by elastic forces. If these atoms having a finite mass are excited by an alternating electromagnetic field, resonance will occur when the stimulating frequency equals that corresponding to the difference between two quantized vibrational energy levels. The vibrational spectra of importance all lie in the infrared range of frequencies and thus outside the scope of this treatment.

It then only remains to consider one further type of molecular spectrum, namely that originating from paramagnetic resonances.

Paramagnetic Spectra. These spectra are of great importance in connection with atmospheric absorption. Let us now study the group of molecules which contain one or more unbalanced electron spins in the electronic ground state. The molecules O_2, NO, NO_2 and ClO_2 are examples. If the molecule is subjected to a magnetic field, the unbalanced spins will be weakly coupled to the magnetic momentum resulting from this field. Now, such a magnetic field will under certain circumstances be generated from the orbital motion of the electrons or from the end-over-end rotation of the whole molecule. This weak interaction between the spin magnetic moment S and the magnetic moment generated by the molecular rotation results in a splitting of each rotational level into $(2S + 1)$ components, commonly known as the spin multiplets. Transitions between these triplets, brought about by an external magnetic field of frequency ν, will result in closely spaced absorption lines for O_2 in the 4 to 6 mm region and to a single component in the 2.5 mm region [63].

Having now discussed the problem phenomenologically, let us give some quantitative results. The absorption coefficient at the peak resonance frequency is given by:

$$\alpha_o = \frac{8\pi^2 \nu_0^2 Nf}{3CkT} (\mu_{ij}) \frac{1}{\Delta \nu} \qquad (5.75)$$

where N   = total number of molecules per ml
           f   = fraction of the total molecules present in the lower state
           μ_{ij} = matrix element of the dipole moment μ, which gives rise to transition from i to j states
           $\Delta \nu$ = half-width of absorption line
           ν_0 = peak resonance frequency

In terms of the conventional measure of attenuation dB/km, we have:

$$\gamma = 10^6 \alpha_\nu \log_{10} e \text{ dB/km} \qquad (5.76)$$

Liquid and Solid Particles

When dealing with this part of the problem, the procedure is to determine by exact electromagnetic theory the power absorbed and lost by scattering when a plane wave passes over a single spherical droplet. The result is dependent only on the wavelength λ, the drop diameter D and its complex dielectric constant. The dielectric constant is a function of λ and also of the temperature T.

The effect of a uniform concentration of droplets, all of the same size, may be determined, and from this information the attenuation can be derived for any droplet concentration and size distribution applicable to particular meteorological conditions. The scattering particles are assumed to be distributed at random, and the scattered radiation is considered as entirely noncoherent for the range of wavelengths under consideration. The effects of the passage of a plane wave over a sphere have been studied intensively in the past, generally with various restrictive conditions. The completely general and exact theory was first developed by Mie in 1908, and a compact presentation of the problem was given by Stratton in 1941 [4]. On the basis of these earlier works, the effects produced by single drops of various diameters can be determined. The only fundamental physical information required for this is a knowledge of the complex dielectric constant ϵ of the drop for the appropriate ranges of λ and T.

$$\epsilon = \epsilon' - j\epsilon''$$

The square root m of the dielectric constant may be written

$$m = n - j\, n\psi \qquad (5.77)$$

where n = refractive index
 $n\psi$ = absorption index

It is on the values of these two parameters, for a given λ and T, that the required absorption and scattering properties of a sphere of a given size will depend.

130 ADAPTIVE RADAR

Now let us consider the absorption effects of the liquid particles. On the basis of Mie and Stratton's work, it can be shown that the attenuation in dB/km is given by:

$$\alpha(\mathrm{dB/km}) = 10 \log_{10} e \cdot 10^5 \cdot N \frac{\pi D^2}{4} f_a \left(\frac{D}{\lambda}, m\right) \quad (5.78)$$

where N = number of drops
       D = diameter of drop
       f_a = function of D/λ and m

f_a may be regarded as the ratio of the total energy absorbed and scattered by a single drop, to that in the area of the wavefront equal to the projected area of the drop. In fact, the effective target-area of a sphere for combined absorption and total scattering is $(\pi D^2/4)f_a$.

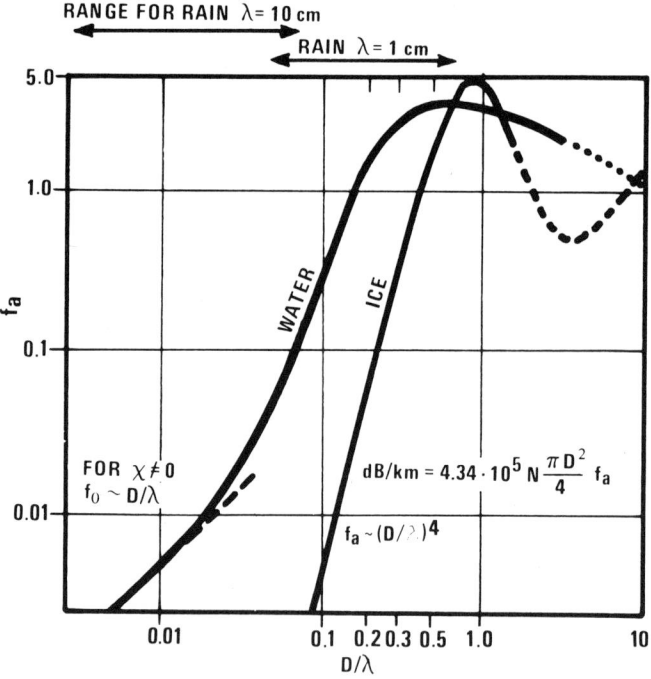

Figure 5.31 Effective cross section of a sphere for combined absorption and total scattering $\pi D^2/4f_a$. Schematic diagram: n and nψ assumed constant at values slightly greater than 1 cm (after Ryde [64]).

The determination of f_a is simple when the drop diameter is very small in comparison with the wavelength. However, as D/λ exceeds about 0.06, the computations rapidly become lengthy, because Riccati-Bessel and Hankel functions with complex arguments are involved. The general way in which f_a is found to vary with D/λ is shown in Figure 5.31 for a constant value of m appropriate to a wavelength slightly greater than 1 cm and T = 18°C.

Now let us consider the limiting case of fine-droplet clouds where D < 0.01 cm. From Figure 5.31 we may see that, provided D/λ < 0.015, f_a varies as D/λ. It may, in fact, be shown that for such small values of D/λ the function f_a reduces to:

$$f_a = 2\pi C_1 D/\lambda \qquad (5.79)$$

where C_1 is a function of n and $n\psi$ and therefore also of λ. The expression for the attenuation then becomes:

$$\alpha(dB/km) = 4.093\, C_1\, M/\lambda\rho \qquad (5.80)$$

where ρ = density
M = mass concentration of droplet (g/m³)

Note that the attenuation does not depend on drop size D.

In this brief presentation, it does not appear to be pertinent to go into detail about the various types of liquid and solid particles in the troposphere. The attenuation produced by hail is calculated very much in the same way as that for rain droplets. As far as the absorption caused by snow and ice-clouds is concerned, Mie's theory developed specifically for spheres will no longer apply. The ice crystals appear in the form of stars, plates, prisms and prismatic needles, according to the conditions under which they are formed. Gans in 1912 considered the case of ellipsoids, small in comparison with the wavelength, and from this work expressions relating to the limiting case of small disks and needles were derived.

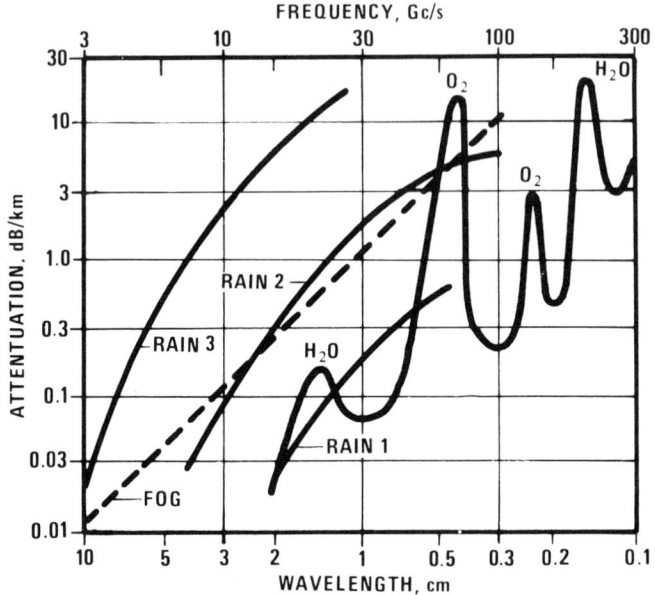

Figure 5.32 Tropospheric absorption (after du Castel [65]). Rain 1 = 1 mm/hr, rain 2 = 10 mm/hr, rain 3 = 100 mm/hr, fog = 25-m visibility.

It was found that for ice crystals, whose principal dimensions do not normally exceed a few millimeters, expression of the form established above for fine water droplets (Equation 5.80) will be sufficiently accurate for wavelength in the cm band. It is found that even for $\lambda = 1$ cm the attenuation is extremely small.

Figure 5.31 summarizes the absorbing effect on radio waves of gases, solid and liquid particles in the troposphere. We see that the gases produce a very frequency-selective absorption, the general trend being an increase in absorption with increasing frequency. The solid and liquid particles in the troposphere produce a smooth curve relationship between attenuation and frequency; high frequency is associated with high attenuation [66,67].

CHAPTER 6

"INTELLIGENT" RADAR: ADAPTIVE RADAR SYSTEMS IN RELATION TO TARGET, TERRESTRIAL BACKGROUND AND PROPAGATION MEDIUM

We introduced the book by suggesting that most of the existing detection/identification systems hitherto available do not make optimum use of all the a priori information on the object of interest of which one is generally in possession. It was suggested that an optimum radar system can be designed if we have sufficient information about the shape of the object, its motion pattern and the terrestrial background against which the target is viewed. Finally, it was emphasized that the transmission medium between the observation platform and the target introduces multiplicative noise (distortion) which may impose severe limitations in regard to the detection/identification potential of our radar system.

We shall end this book by substantiating these statements in the form of a synthesis of the intervening chapters. Figure 6.1 may serve as an illustration to this final chapter. The aim is to detect and identify a target against a background through a more or less adverse propagation medium. We shall process the illuminating wave as well as the received signal in such a way so as to optimize the overall system performance.

As an example, let us consider an over-the-horizon radar with a view to identify ships on the sea-surface. As we have seen (Chapter 3), to resolve a target of a given size, we shall have to

134 ADAPTIVE RADAR

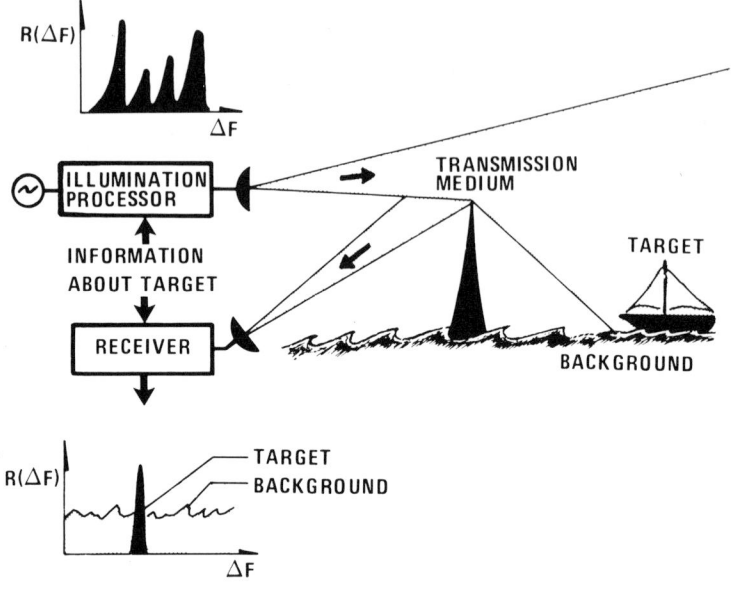

Figure 6.1 We are faced with the consideration of three "filter functions": the transmission medium between the observer and the target, the background, and the target itself.

illuminate the target with a set of electromagnetic waves the frequency spacing of which must be larger than $c/2\Delta z$ if Δz is the longitudinal size of the object, and the wavefront must be such that the transverse correlation distance of the electromagnetic field is larger than the transverse size Δx and Δy of the target. With these conditions in mind, let us consider the troposcatter transmission medium.

We have previously shown (Chapter 5) that the troposcatter bandwidth and the spatial correlation properties of fieldstrength vary substantially with the conditions prevailing. Specifically, Figure 5.13 shows that a bandwidth of approximately 1.4 MHz is exceeded 1% of the time, whereas a bandwidth of 0.6 MHz is exceeded in more than 50% of the time, when the transmission path is 300 km long. These bandwidth limitations are illustrated in Figure 6.2 with reference to the "bandwidth signatures" of

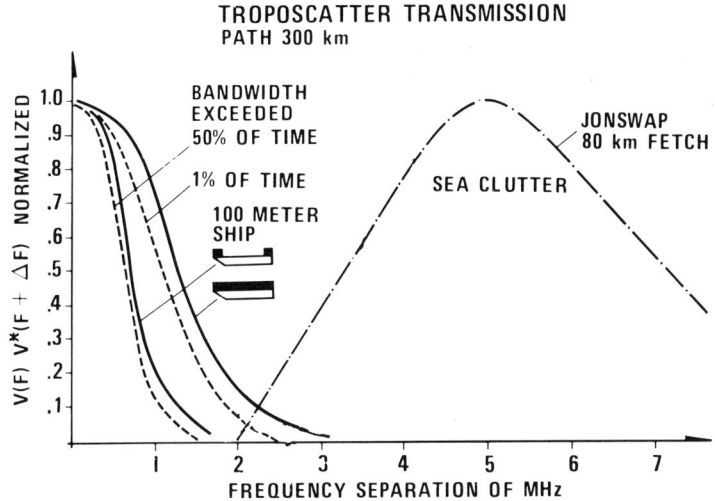

Figure 6.2 Design considerations in relation to an over-the-horizon troposcatter radar for detection and classification of ships against a sea-clutter background.

two different types of 100-m ships. Note that a 100-m ship with the scatterers confined to the bow and the stern, respectively, can be resolved 1% of the time, whereas a 100-m ship with an even distribution of the scatterers cannot be resolved 1% of the time.

Note that the contribution from the sea-clutter is of very limited importance since the irregularity spectrum of the sea falls outside that of the target. Accordingly, in order to avoid sea-clutter contributions, the bandwidth of our radar illuminator should be limited to approximately 2 MHz.

Then let us consider what the troposcatter transmission medium does to the spatial coherence properties of the electromagnetic waves illuminating the target. These correlation properties were discussed in Chapter 5, and summarized in Figures 5.14 and 5.15. We see that for a 300-km troposcatter path a horizontal fieldstrength correlation distance of approximately 30 wavelengths is obtained in 1% of the time, whereas a correlation distance of approximately 20 wavelengths is exceeded in 50% of the time. This means that if we were to investigate the transverse horizontal distribution of scatterers constituting the target using an antenna configuration as that depicted in Figure 2.7 and 10-cm radar, we should not be able to illuminate a target larger than 3 m in a coherent manner. If, on the other hand, we are basing our investigation on a vertical antenna array, thus measuring the vertical distribution of the target scatterers, we shall be able to illuminate a 5-m target in a coherent manner using a 10-cm radar system.

In summing up this section on the multifrequency, over-the-horizon radar system, note that Figure 6.2 relates the JONSWAP ocean wave spectrum predictions to results from theoretical radar target investigations.

Having, in broad terms, considered the potential of a multifrequency adaptive radar system in relation to targets on the sea surface, we shall now give an example where a target signature is obtained both theoretically and on the basis of experiments. In Chapter 2 we calculated the multifrequency radar signature from knowledge of the spatial distribution of scatterers within the target. As an illustration of such calculations, Figure 2.3 shows the signatures related to different classes of targets. Let us assume that an airplane from nose to tail can be approximated by a Gaussian distribution of scatterers. Inspecting an aircraft such as the F-16 fighter, it is obvious that contributions to the total scatterering cross section from the nose and from the tail is small. The dominating contributions will probably be stemming from the region in the vicinity of the wings.

In Figure 6.3, the results are presented of theory as well as experiments [61]. The experimental results were obtained using an F-16 aircraft flying close to the sea surface toward the radar. Note that the sea-clutter contributes in the low frequency end of the spectrum only.

So far we have been considering target and background signatures in one spatial domain only. We have already shown (see Figures 4.3 and 4.4) that a plane ocean wave scatters the radio waves back within a rather narrow range of azimuth

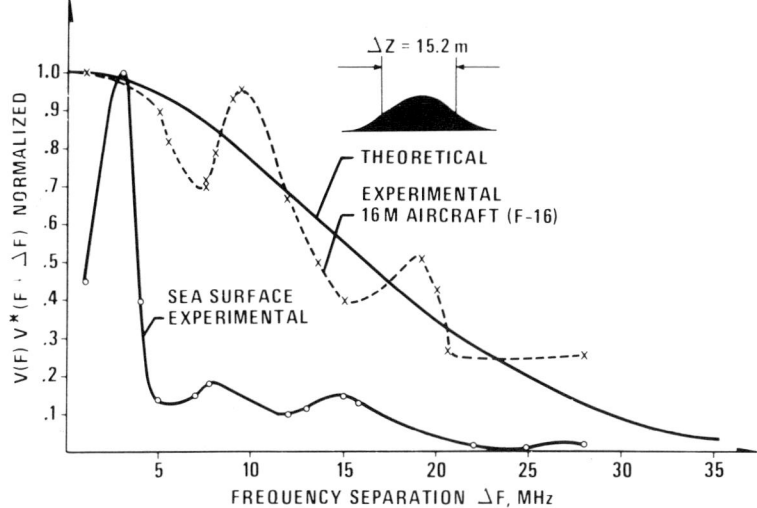

Figure 6.3 Longitudinal distribution of scattering centers for F-16 aircraft. There are a few dominant scattering centers and, apparently, many smaller ones that are distributed along the main axes of the aircraft. The sea-surface contributions are limited to the low-frequency end of the spectrum.

angles. Hence, to increase the contrast of a target against a sea-clutter background, the target should be illuminated within a range of azimuth angles which does not coincide with that giving maximum backscatter from the sea surface.

The "directivity" of the sea surface is shown in Figure 6.4. We have plotted Equation 4.4 to the basis of ocean wavelength using the distance between the radar and the scattering surface R as the parameter. Note that the longer the ocean wavelength is, the wider is the scattered beam. Note also that a large range R gives rise to a narrow scattered beam and conversely a short range results in a wide beam. Note, however, that the azimuth sector within which a radar antenna with diameter d is receiving backscattered energy cannot be smaller than that determined by the antenna apparature:

$$\text{beamwidth } \beta = \frac{\text{radio wavelength } \lambda}{\text{antenna apparature } d}$$

138 ADAPTIVE RADAR

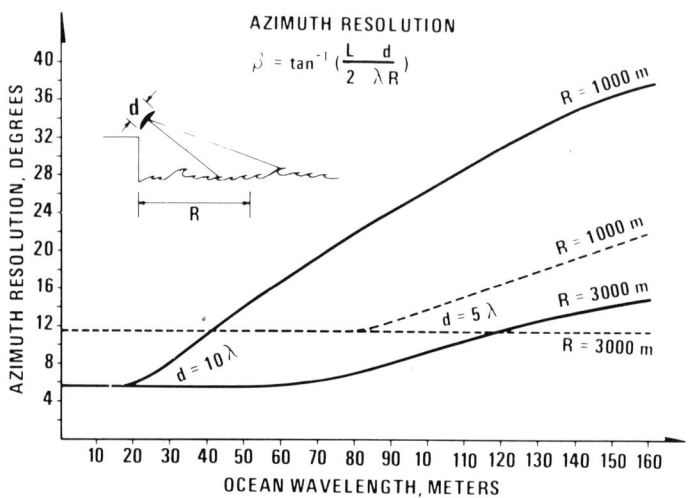

Figure 6.4 Ocean waves scatter radiowaves back within a small-azimuth sector. The larger the ocean wavelength, the wider is the azimuthal spread of the backscattered waves. The wider the illuminated area on the sea surface, the narrower is the beamwidth of the backscattered waves.

Note that the theoretical curves of Figure 6.4 refer to the very simple case where the phase front of the ocean wave is plane. In practice, the irregularity structure of the sea surface may be the result of several wave systems with different propagation direction. Under such conditions one would not expect to find an azimuthal dependence as that depicted in Figure 6.4. Experimental results [31], however, indicate in many cases that the simple relationship of Figure 6.4 is valid. An example of such an azimuthal distribution is shown in Figure 6.5

Before we conclude this section on signature matching techniques, we shall include another signature domain: the temporal (Doppler) signature. As we have seen in earlier chapters, the target, and also the background, can be characterized by a motion pattern in addition to a geometric size and shape. In

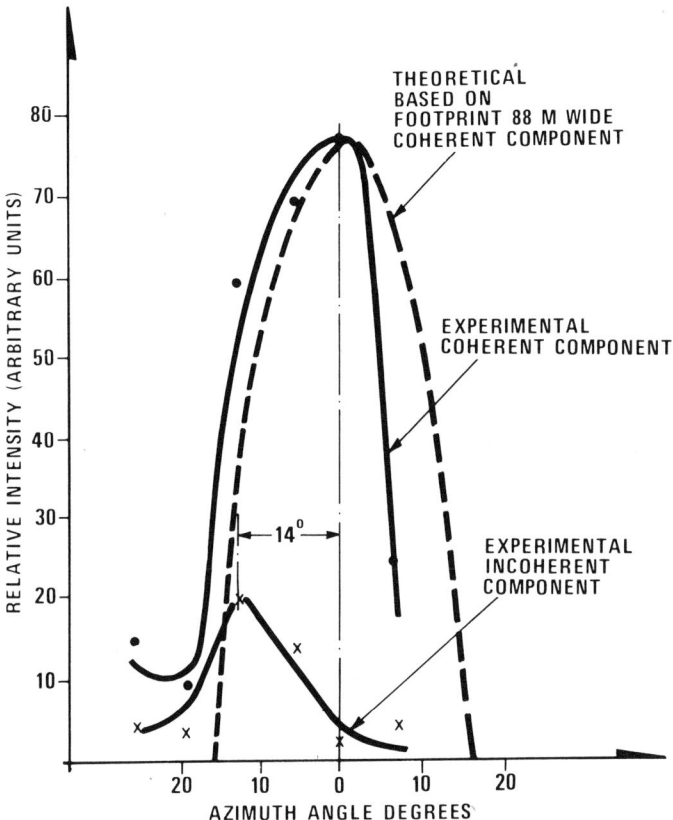

Figure 6.5 The angular distribution of an ocean wave of wavelength 50 m. The solid line gives the direction distribution for the "coherent" gravity wave; the curve marked with Xs gives the distribution of the "incoherent component" [68]. This incoherent distribution is shifted some 14° relative to the coherent wave (see Figures 4.3 and 4.4).

Chapter 4 we established the relationship between Doppler shift f and frequency difference ΔF between two illuminating waves:

$$f_s = \sqrt{\frac{g \Delta F}{\pi c}}$$

140 ADAPTIVE RADAR

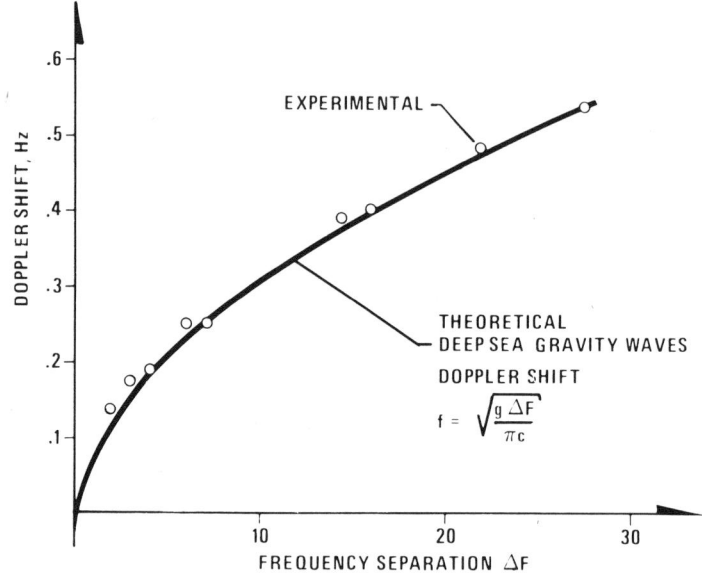

Figure 6.6 Deep-sea gravity waves are dispersive: there is a square-root relationship between phase velocity and wavelength. The experimental points lie very close to the theoretical curve. The discrepancy may be caused by an ocean current [31,68].

This dispersion relationship for deep sea gravity waves was plotted in Figure 4.5. Experimental verification is provided in Figure 6.6. Note that the experimental points lie very close to the theoretical curve, but they are shifted systematically a small amount toward higher Doppler frequencies. This is probably due to a steady current superimposed on which the gravity waves are propagating.

Then, let us consider the Doppler signature of a rigid body moving along a straight line toward the radar with a constant velocity. Since, in this case, all the scatterers to which our multifrequency radar is matched move at the same velocity, being an integral part of the same body, a linear relationship should exist between Doppler shift and illuminating frequency.

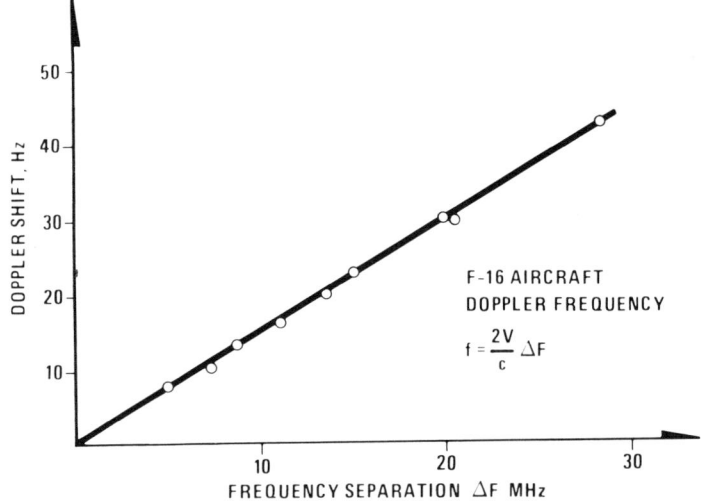

Figure 6.7 A rigid body moving at constant velocity toward the radar gives a Doppler shift that is proportional to the illuminating frequency and the velocity of the object.

This was previously discussed in Chapter 4 and illustrated in Figure 4.5. The Doppler shift of the target is given by:

$$f_T = \frac{\Delta F}{c} \cdot 2V$$

This linear (nondispersive) relationship is, not surprisingly, confirmed experimentally in Figure 6.7.

As depicted in Figure 4.5, target discrimination is achieved by Doppler filtering unless the target velocity is such that the Doppler spectrum stemming from the target coincides with that from the sea surface. This is the case when

$$\frac{\Delta F}{c} \cdot 2V = \sqrt{\frac{g \Delta F}{\pi c}}$$

i.e., when

$$\Delta F = \frac{g}{\pi} \frac{c}{4V^2}$$

To distinguish between target and sea-clutter under such conditions, we can investigate the slope of the Doppler vs ΔF relationship. As depicted in Figure 4.5, when the Doppler shifts coincide, the first derivatives are not equal. The slope related to the target is given by:

$$\left(\frac{df}{d\Delta F}\right)_T = \frac{2V}{c}$$

The corresponding slope for the dispersive gravity waves is similarly given by:

$$\left(\frac{df}{d\Delta F}\right)_S = \frac{g}{2\pi c} \sqrt{\frac{c}{g \Delta F}}$$

Hence, when the Doppler shift resulting from sea-clutter is identical to that resulting from the motion of the target then the ratio of the two gradients is given by:

$$\left(\frac{df}{d\Delta F}\right)_S \left(\frac{df}{d\Delta F}\right)_T^{-1} = \frac{1}{2}$$

When, on the other hand, the two gradients are the same, i.e., when

$$\left(\frac{df}{d\Delta F}\right)_S = \left(\frac{df}{d\Delta F}\right)_T$$

then the ratio of the Dopper shift

$$\frac{f_s}{f_T} = 2$$

In conclusion, it should be noted that the target can be discriminated from the sea-surface even when the two respective Doppler shifts are equal by inspecting the first derivatives.

REFERENCES

1. Gjessing, D. T. *Remote Surveillance by Electromagnetic Waves for Air - Water - Land* (Ann Arbor, MI: Ann Arbor Science Publishers, Inc., 1978).
2. Gabriel, W. F. "Adaptive Arrays–An Introduction," *IEEE Proc.* 64: (2) (1976).
3. Applebaum, S. "Adaptive Arrays," *IEEE Trans. Antennas Propagation* AP-24(5):585-589 (1976).
4. Stratton, J. A. *Electromagnetic Theory* (New York: McGraw-Hill Book Company, 1941).
5. Megaw, E. C. S. "Fundamental Radio Scatter Propagation Theory," Proc. IEEE, Monog. No. 236R (1957).
6. Batchelor, G. K. "The Scattering of Radio Waves in the Atmosphere by Turbulent Fluctuations in Refractive Index," Research Report No. EE262, School of Engineering, Cornell University (1955).
7. Bass, F. B., and I. M. Fuks. *Wave Scattering from Statistically Rough Surfaces* (Elmsford, NY: Pergamon Press, Inc., 1979).
8. Wiener, N. "Generalized Harmonic Analysis," *Acta Math.* Vol. 55 (1930).
9. Gjessing, D. T. "Environmental Remote Sensing, Part I: Methods Based on Scattering and Diffraction of Radio Waves," *Phys. Technol.* Vol. 10 (1979).
10. Gjessing, D. T. "Atmospheric Structure Deduced from the Forward Scatter Wave Propagation Experiments," *Radio Sci.* 4(12):1195-1210 (1969).
11. Gjessing, D. T. "Target Detection and Identification Methods Based on Radio- and Optical Waves," AGARD Lecture Series No. 93 (1978).

12. Gjessing, D. T. "A Generalized Method for Environmental Surveillance by Remote Probing," *J. Radio Sci.* (March/April 1978).
13. Nathanson, R. D. *Radar Design Principles* (New York: McGraw-Hill Book Company, 1969).
14. Weissman, D. E., and J. W. Johnson. "Dual Frequency Correlation Radar Measurements of the Height Statistics of Ocean Waves," *IEEE J. Ocean. Eng.* OE-2:74-83 (1977).
15. Tomiyasu, K. "Short Pulse Wide-Band Scatterometer Ocean Surface Signature," *IEEE Trans. Geosci. Electron.* GE-9:175-177 (1971).
16. Weissman, D. E., and J. W. Johnson. "Rough Surface Wavelength Measurement through Self Mixing of Doppler Microwave Backscatter," *IEEE Trans. Antennas Propagation* AP-27:730-737 (1979).
17. Clifford, S. F., and D. E. Barrick. "Remote Sensing of Sea State by Analysis of Backscattered Microwave Phase Fluctuations," *IEEE Trans. Antennas Propagation* AP-26:699-705 (1978).
18. Barrick, D. E. "First-Order Theory and Analysis of MF/HF&VHF Scatter from the Sea," *IEEE Trans. Antennas Propagat.* AP-20:2-10 (1972).
19. Plant, W. J. "Studies of Backscattered Sea Return with a CW Dual-Frequency X-Band Radar," *IEEE Trans. Antennas Propagation* AP-25(1) (1977).
20. Schuler, D. L. "Remote Sensing of Directional Gravity Wave Spectra and Surface Currents Using a Microwave Dual-Frequency Radar," *Radio Sci.* 13(2) (1978).
21. Valenzuela, G. R., et al. "Modulation of Short Gravity-Capillary Waves by Longer Scale Periodic Flows," Technical Note, Naval Research Laboratory.
22. Valenzuela, G. R. In: *Proceedings of the NATO Advanced Study Institute*, T. Lund, Ed. (1978).
23. Gjessing, D. T., and R. Irgens. "On the Scattering of Electromagnetic Waves by a Moving Tropospheric Layer Having Sinusoidal Boundaries," *IEEE Trans. Antennas Propagation* AP-12 (1964).
24. Gjessing, D. T. "Adaptive Techniques for Radar Detection and Identification of Objects in an Ocean Environment," *J. Ocean Eng.* OE-1(1) (1981).
25. Sittrop, H. Personal communication (1981).
    of Sea-Clutter, Parts I and II," Reports PHL-1975-08 and -09, Physics Laboratory TNO (1975).
26. Ramamonjiarisoa, A., et al. "Observations de la vitesse de propagation des vagues engindrees par let vent au large," *C. R. Acad. Sci., Paris* (September 18, 1978), p. 287.

REFERENCES 147

27. Ramamonjiarisoa, A., et al. "Laboratory Studies on Wind-Wave Generation, Amplification and Evolution," in *Turbulent Fluxes through the Sea-Surface: Wave Dynamics and Prediction*, A. Faure and K. Hasselman, Eds. (New York: Plenum Publishing Corporation, 1978).
28. Phillips, O. M. *The Dynamics of the Upper Ocean* (Cambridge, England: Cambridge University Press, 1977).
29. Gjessing, D. T. "On the Scattering of Electromagnetic Waves by Nonisotropic Inhomogeneities in the Atmosphere," *J. Geophys. Res.* 67(3) (1962).
30. Hasselmann, K., et al. "Measurements of Wind-Wave Growth and Swell Decay During the Joint North Sea Wave Project (JONSWAP)," *Erganzungsheft zur Deutschen Hydrographicschen Zeits.* Reihe A/80 (12) (1973).
31. Gjessing, D. T., J. Hjelmstad, A. G. Kjelaas and T. Lund. "Directional Ocean Wave Spectra as Observed with a Multifrequency Radar System" (in preparation.)
32. Bean, B. R., and E. J. Dutton. "Radio Meteorology," U.S. Dept. of Commerce, National Bureau of Standards, Monog. 92, U.S. Government Printing Office (1966).
33. Lee, R. W., and J. C. Harp. "Weak Scattering in Random Media," *J. IEEE* 57:375 (1969).
34. Gjessing, D. T., A. G. Kjelaas and J. Nordo. "Spectral Measurements and Atmospheric Stability," *J. Atmos. Sci.* 26(3):462.8 (1969).
35. Gjessing, D. T., and K. S. McCormick. "On the Prediction of the Characteristic Parameters of Long-Distance Tropospheric Communication Links," *IEEE Trans. Comm.* 22(9) (1974).
36. Gjessing, D. T. "On the Use of Forward Scatter Techniques in the Study of Turbulent Stratified Layers in the Troposphere," *Boundary Layer Meteorol.* 4:377-396 (1973).
37. Gjessing, D. T. "An Experimental Determination of the Spectrum of Permittivity and Air Velocity Fluctuations Along a Vertical Direction in the Troposphere Using Radio Propagation Methods," *J. Atmos. Terr. Phys.* 26:2 (1964).
38. Bull, G., and J. Neissner. "Untersuchung der atmosfärischen Feinstruktur mit Hilfe von Ausbreitungsmessungen in Mikrowellenbereich," *Beitr. Geophys.* 77(5):394-410 (1968).
39. Eklund, F., and S. Wicherts. "Wavelength Dependence of Microwave Propagation Far Beyond the Radio Horizon," *Radio Sci.* 3(11):1068-1974 (1968).

40. Fehlaber, L. "Diversity Abstand auf Scatter Strecken im Frequenzbereich zwischen 1 GHz and 10 GHz," *Tech. Vericht.* 5581 (1966).
41. Fehlaber, L., and J. Grosskopf. "Messung der Gewinnminderung bei 1715 MHz aur einer 409 km langen Scatter-Versuchsstrecke," Fernmeldetechnisches Zentralamt FTZ-A455 (1968).
42. Bolgiano, R., Jr. "A Study of Wavelength Dependence of Transhorizon Radio Propagation," Research Report 188, Cornell University (1964).
43. Cox, D. C., and A. T. Watermann, Jr. "Phase and Amplitude Measurements of Transhorizon Microwaves with a Multidata-Gathering Antenna Array," AGARD Conference Proceedings Vol. 37, pp. 18-1 to 18-6 (1968).
44. Gjessing, D. T. "Scattering of Radio Waves from Regular and Irregular Time Varying Refractive-Index Structures in the Troposphere," AGARD Conference Proceedings, Vol. 37, pp. 15-1 to 15-17 (1968).
45. Gjessing, D. T., and J. A. Borresen. "The Influence of an Irregular Refractive-Index Structure on the Spatial Field-Strength Correlation of a Scattered Radio Wave, IEEE Conference Publ. 48, pp. 43-50 (1968).
46. Gjessing, D. T. "Scattering Mechanisms and Channel Characterization in Relation to Broad-Band Radio Communication Systems," AGARD Conference Proceedings CP-244 (1977).
47. Wait, J. R. "A Note on VHF Reflection from a Tropospheric Layer," *Radio Sci.* 7:847-878 (1964).
48. Grosskopf, J. "Investigation of the Receiving Field for Scatter Propagation," *AGARD Conf. Proc.* 37:22-1 to 22-11 (1968).
49. Hall, M. P. M., Appelton Laboratory, U.K. Personal communication.
50. Lumley, J. L. "Theoretical Aspects of Research on Turbulence in Stratified Flows," in *Atmospheric Turbulence and Radio Wave Propagation*, A. M. Yaglom and V. I. Tatarsky, Eds. (Moscow, U.S.S.R.: Publishing House Nauka, 1967).
51. Waterman, A. T., Jr., D. T. Gjessing and C. L. Liston. "Statistical Analysis of Transmission Data from a Simultaneous Frequency- and Angle Scan Experiment," paper presented at the URSI spring meeting, Washington, DC (1961).
52. McGillem, C. D., C. R. Cooper and W. B. Waltman. "Use of Wideband Stochastic Signals for Measuring Range and Velocity," *Easton 69 Record* (1969), pp. 305-311.
53. Ottersten, H., K. R. Hardy and C. G. Little. "Radar and Solar Probing of Waves and Turbulence in Statically Stable Clear-Air Layers," *Boundary-Layer Meteorol.* (1973), pp. 47-89.

54. Saxton, J. A., J. A. Lane, R. W. Meadows and P. A. Mathews. "Layer Structure of the Troposphere," *Proc. IEEE* 111:2 (1964).
55. Gjessing, D. T. "Radiophysical Aspects of Irregular Structure in the Atmosphere," in *Atmospheric Turbulence and Radio Wave Propagation*, A. M. Yaglom and V. I. Tatarsky, Eds. (Moscow, U.S.S.R.,. Publishing House Nauka, 1967).
56. Seehars, H. D. "Investigation of the Dielectric Turbulence and Wind Stratification in the Troposphere by Means of SHF Beamswinging Experiments over Sea," Report No. 18, Instituts für Radiometeorologie und Maritime Meteorologie, Universitat Hamburg (1970).
57. Gjessing, D. T., H. Jeske and N. Klint-Nansen, "An Investigation of the Tropospheric Fine Scale Properties Using Radio, Radar and Direct Methods," *J. Atmos. Terr. Phys.* 31:1157-1182 (1969).
58. Gjessing, D. T., and F. Irgens. "Scattering of Radio Waves by a Moving Rippled Layer: A Simple Model Experiment," *IEEE Trans. Antennas Propagat.* AG-12:6 (1964).
59. Kerr, D. E. *Propagation of Short Radio Waves* (Dover Publications, 1965).
60. McCormick, G. C., and A. Ivendry. "Depolarization over a Link Due to Rain: Measurement of Parameters," *Radio Sci.* 11:741-749 (1976).
61. Drufuca, G. "Rain Attenuation Statistics for Frequencies above 10 GHz from Rain Gauge Observations," *J. Rech. Atmos.* 8(1-2): 399-411 (1974).
62. Ratcliffe, J. A. "Some Aspects of Diffraction Theory and Their Application to the Ionosphere," *Phys. Soc. Rep. Prog. Phys.* 19: 188 (1965).
63. Gordy, W. "Microwave Spectroscopy," in *Handbook of Physics* (Berlin: Springer-Verlag, 1957).
64. Ryde, J. W. "The Attenuation and Radar Echoes Produced at Centimeter Wavelengths of Various Meteorological Phenomena," The Physical Society, London (1946).
65. Du Castel, F. "Propagation tropospheric et faisceaux hertzienx transhorizon, Collection Technique et Scientifique de CNET," Editions Chiron, Paris (1961).
66. Ghosh, S. N., and V. Malaviya. "Microwave Absorption in the Earth's Atmosphere," *J. Atmos. Terr. Phys.* 21:4 (1961).
67. Burrows, C. R. *Radio Wave Propagation* (New York: Academic Press, Inc., 1949).

68. Gjessing, D. T., J. Hjelmstad and T. Lund. "A Multifrequency Adaptive Radar for Detection and Identification of Objects: Results of Preliminary Experiments on Aircraft against a Sea-Clutter Background" (in press).

INDEX

absorption 126,128,131
amplitude covariance 60
amplitude spectrum 16
angle of arrival spectrum 62,75,96,
   125
angle of incidence 36,113
angular distribution 124
angular field distribution 19
angular power distribution 20,21
angular power spectrum 62,69,75,
   88,93,96,101,106,121
angular spectrum 20
antenna
   aperture 20,42,63,77
   array 3,21,22,136
   elements 19
   gain 76,77,97
   -to-medium coupling loss 69,77
aspect angle 28
atmospheric gases 127
atmospheric layer 92
atmospheric refraction 96
atmospheric structure 98
atmospheric turbulence 64
atomic spectra 126
autocorrelation 10,12,16,20,29
   complex 74
   spatial 35,66,68
averaging schemes 10,35

background signatures 3,29,136
bandpass filter 118
bandwidth 9,18,35,53,57,69,71,
   78,80,98,101,102,115,116,
   122
   instantaneous 65
   signatures 134
beam geometry 95
beamswinging experiments 86,98
beamwidth 121
boundary layer 36,38
Bragg conditions 5
Bragg scattering 41,111

capillary waves 36,38,39
charge-coupled devices 1
coherence 9,29,49,64,135
   spatial 53,64
   filtering 52
convolution 13,107
   integral 93
corner reflector 29
correlation 11,16,18,75,97,102,
   123,125,134,135
   complex 75,96
   distance 63,69,74,78,80
   spatial 75,96,123
coupling loss 42,78,102
covariance function 46,49,122

cross-beam drift 25
cross-correlation function 72
cross spectrum coherence 52

damped oscillations 124
damping factor 111,113,114
damping function 117,124
Debye relationship 81
delay
    function 8,10,11,23,28,29,
        30,35,36,87,116,119
    loci of constant 62,98,99
    spectrum 69,70,78,80,100,
        101
dielectric constant 104,129
dielectric sphere 103,104
diffracted power spectrum 114
diffracted wave 107,121,122
diffracting angles 105
diffraction 36,67,106,108,113,
    124,125
    grating 117
dipole moment 6,66,103,128
dipole resonance 104
dispersion 34,37,40,46,140,142
Doppler broadening 24,25,31,47
Doppler filtering 31
Doppler shift 23,31,32,46,49,141
Doppler signature 45,138
Doppler spectrum 24,43

earth radius 59,81,97,99
elastic forces 127
electronic ground states 127

fading 65
field-strength distribution 109,
    123
filter, spatial 22
focusing effect 36
Fourier analysis 16

Fresnel 105
Fresnel-Kirchhoff 105
Fresnel zone 62,63

gravity waves 37,38,40,44,139,
    140

Hankel function 131
Huygen's principle 105

ice crystals 131
identification 52
illumination function 30
impulse response 73
interference 56
inversion spectra 127
irregularity scale 23
isotropy 79

JONSWAP ocean wave spectrum
    136

Kelvin-Helmholtz instabilities
    37
Kirchhoff 105
knife-edge diffraction 104,108,
    111,112,122,123

layered structure 98
line-of-sight propagation 59

magnetic moment 128
matched illumination 3,34,42
matched receiver 1
Maxwell's equations 8,66
meteorological factors 81
microwave transitions 127
Mie scatter 105,129
molecular rotation 128
molecular spectra 127
motion, translatory 49
motion pattern 32,138

multipath effects 59,98,106,109, 112

network theory 73
noise, additive 2
noise, multiplicative 2,133

obstacle gain 112,113,114
ocean waves 36,37,40,43
orbital motion 128
over-the-horizon scatter 65

paramagnetic spectra 128
path length 98,99,100,116
pathloss 53,64,114
pattern recognition 1
permittivity 66,107
phased-array 3,22,136
polarization 56
    potential 6,66
power spectrum 13,68,73
    spatial 69

radar
    antenna 137
    illuminator 3,15
    intelligent 133
    multifrequency 16,22,46
    over-the-horizon 133,135
    signature 27,29,40,49
    target-adaptive 27
radiometeorological parameters 71,80,81,82
radiosonde observations 86
rain 103
raindrops 103,129,131
Rayleigh scatter 104,105
reflection 55,105,109,111,118, 124
    coefficient 90,91,113,119
refraction 57,69,81,82,87,95

refractive index 57,69,81,82,84, 85
    irregularity spectrum 75,77, 81,93,95,99
Riccati-Bessel function 131
Richardson's number 84
rotational spectra 127
rough surface 16

scatter, forward 81,98
scatterers, distribution 18,28,137
scattering
    angle 66,82
    centers 24
    coefficient 36
    cross section 8,18,29,37,79, 104
    mechanisms 67,69,79,95,98
    particles 103
    sea-surface 37
    volume 67,71
sea
    -clutter 135,136,142
    index 37
    state 37
    -surface 15
shadowing effects 36
sidebands 72
signature domains 29
signature matching techniques 138
single-layer structure 98
slope spectrum 79,81
Snell's law 57,81
Sommerfeld 105
spatial spectrum 23,40,97
specific heat 84
spin 128
spread spectrum modulation 65
surface structure, irregular 36, 45,138

target parameters 27,28,29,30, 130,141
target, transverse structure of 18
terrain obstacle 114
transfer function 73
transmission loss 96,108,122
transmission medium 3,123
troposcatter 134,135
turbulence 38,44,91,94,95,100, 101

turbulent atmosphere 98

Väisälä-Brunt frequency 84
velocity shear 37
velocity spread 24,47
vibrational spectra 127
velocity spread 24,47
vibrational spectra 127

water vapor pressure 81